Video Segmentation and Its Applications

King Ngi Ngan · Hongliang Li

Editors

Video Segmentation and Its Applications

Springer

Editors
King Ngi Ngan
Department of Electronic Engineering
The Chinese University of Hong Kong
Shatin, New Territories
Hong Kong SAR
China, People's Republic
knngan@ee.cuhk.edu.hk

Hongliang Li
School of Electronic Engineering
University of Electronic Science
  and Technology of China
Chengdu
China, People's Republic
hlli@uestc.edu.cn

ISBN 978-1-4899-9466-0   ISBN 978-1-4419-9482-0 (eBook)
DOI 10.1007/978-1-4419-9482-0
Springer New York Dordrecht Heidelberg London

Printed on acid-free paper

Springer is part of Springer Science+Business Media (www.springer.com)

Preface

Video segmentation has been a key technique for visual information extraction and plays an important role in digital video processing, pattern recognition, and computer vision. A wide range of video-based applications will benefit from advances in video segmentation including security and surveillance, bank transactions monitoring, video conferencing, and personal entertainment.

In the last four decades, this field has experienced significant growth and progress, resulting in a virtual explosion of published information. The field of image and video segmentation is still a very hot topic, with much advancement in recent years. As a consequence, there is a considerable need for books like this one, which attempts to bring together a selection of the latest results from researchers involved in state-of-the-art work in video segmentation and its applications.

The objective of this book is to present the latest advances in video segmentation and analysis techniques covering both theoretical approaches and real applications. This book provides an overview of emerging new approaches to video segmentation and promising methods being developed in the computer vision and video analysis community. It not only deals with the theoretical foundations and algorithms for image/video segmentation, which includes how to extract video features, and how to segment semantic video objects, this book also provides a comprehensive description of practical applications which I believe fills a hole in the video segmentation market.

This book is expected to provide researchers and practitioners a comprehensive understanding of the start-of-the-art of video segmentation techniques and a resource for potential applications and successful practice. The principal audience of this book will be mainly composed of researchers and engineers as well as graduate students working on video segmentation in various disciplines, e.g. video analysis, computer vision, pattern recognition, image and video processing, artificial intelligence, etc.

Chapter 1 introduces the current status of research activities including graph-based, density estimator-based and temporal-based segmentation algorithms. Recent developments are then discussed while providing a comprehensive introduction to the fields of image/video segmentation. More challenges ahead are identified whilst outlining perspectives for the years to come.

Chapter 2 presents object segmentation algorithms depending on the characteristics of eigen-structure. The eigen-subspaces are obtained from eigen-decomposition of the covariance matrix, which is computed from the selected color samples. By a joint consideration of signal and noise subspace projections of desired colors, the separate eigen-based fuzzy C-means and coupled eigen-based fuzzy C-means are used to achieve effective color object segmentation. With these proposed algorithms, the color objects can be successfully extracted by using eigen-subspace projections.

Chapter 3 addresses the issue of semantic object segmentation, which aims to label each pixel in a video frame to one of the object classes with semantic meanings. An overview of different technologies and major challenges of the semantic object are first discussed for each step. The frameworks of conditional random fields and topic models, which are the representative models of the generative and discriminative approaches respectively, are applied to achieve semantic object segmentation.

Chapter 4 presents a survey and tutorial on the research on the learning-based video-scene analysis. Two major tasks based on their application setup and learning targets are addressed, namely generic methods and genre-specific analysis techniques. Some research challenges in video content analysis and retrieval are reported for the video scene analysis.

Chapter 5 describes the representative and state-of-the-art approaches in multi-view image segmentation and video tracking. A depth-based segmentation in the initial frame and feature-based tracking algorithms from multiview video are proposed for both separated and overlapping human objects.

Chapter 6 discusses segmentation applications such as medical imaging, computer-guided surgery, machine vision, object recognition, surveillance, content-based browsing, and augmented reality applications. The expected segmentation quality for a given application depends on the level of granularity and the requirement that is related to shape precision and temporal coherence of the objects. Although, there exists still significant challenge to perform robust and fully automated segmentation that fits generic tasks, a reliable solution can be achieved using suitable attention and model-based information.

Hong Kong SAR, The People's Republic of China   King Ngi Ngan
Chengdu, The People's Republic of China   Hongliang Li
January 2011

Acknowledgements

The editors, King N. Ngan and H. Li, would like to thank all authors of this book for their great contributions and efforts to make this book *Video Segmentation and Its Applications* possible.

Many of our colleagues provided us with valuable assistance during the writing of this edition, which includes valuable materials used in this book, suggestions, feedback, and comments on this book. Their corrections have had a very positive effect on the whole manuscript. We especially wish to thanks all the students at IVIPC lab who have provided immense help with the preparation of the book in LATEX.

Special thanks go to Jar-Ferr Yang, Wen Gao, E. Izquierdo, and Xiaogang Wang for supporting the production of this book.

Contents

Contributors

Lingyu Duan School of EE & CS, Peking University, Beijing 100871, China

Wen Gao School of EE & CS, Peking University, Beijing 100871, China, wgao@pku.edu.cn

Shu-Sheng Hao Department of Electrical and Electronics Engineering, National Defense University, Tahsi, Taoyuan, Taiwan

E. Izquierdo Department of Electronic Engineering, Queen Mary, University of London, London, UK, ebroul.izquierdo@elec.qmul.ac.uk

Hongliang Li School of Electronic Engineering, University of Electronic Science and Technology of China, Chengdu, China, hlli@uestc.edu.cn

Jia Li Key Lab of Intelligent Information Processing, Institute of Computing Technology, Chinese Academy of Sciences, Beijing 100190, China

Yuanning Li Key Lab of Intelligent Information Processing, Institute of Computing Technology, Chinese Academy of Sciences, Beijing 100190, China

King Ngi Ngan Department of Electronic Engineering, The Chinese University of Hong Kong, Hong Kong, China, knngan@ee.cuhk.edu.hk

Yonghong Tian School of EE & CS, Peking University, Beijing 100871, China

K. Vaiapury Department of Electronic Engineering, Queen Mary, University of London, London, UK

Xiaogang Wang Department of Electronic Engineering, The Chinese University of Hong Kong, Hong Kong, China, xgwang@ee.cuhk.edu.hk

Jar-Ferr Yang Department of Electrical Engineering, National Cheng Kung University, Tainan, Taiwan, jfyang@ee.ncku.edu.tw

Qian Zhang Department of Electronic Engineering, The Chinese University of Hong Kong, Hong Kong, China

Acronyms

AESS	Adaptive eigen-subspace segmentation
CEFCM	Coupled eigen-based fuzzy C-means
CML	Correlations multilabeling
CRF	Conditional random field
DoG	Difference-of-Gaussian
EM	Expectation Maximization
FCM	Fuzzy C-means
FQS	Four quadrant search
FTA	Frequency tuned saliency
GLOH	Gradient Location and Orientation Histogram
GMM	Gaussian Mixture Model
HCRF	Hidden conditional random field
HDP	Hierarchical Dirichlet Process
HMM	Hidden Markov Model
HOG	Histogram of Oriented Gradients
IHO	Integral Histogram of Oriented Gradients
IML	Individual multilabeling
IML-T	Temporal refinement over individual multilabeling
ISVT	Image segmentation and video tracking
KDA	Kernel Discriminant Analysis
LDA	Latent Dirichlet Allocation
LoG	Laplacian of Gaussians
MAP	Maximum a posteriori
MRF	Markov Random Field
MSER	Maximally Stable Extremal Regions
MVI/V	Multiview image/video
MVIs	Multiview images
OCR	Optical character recognition
OOIs	Object-of-interests
PCT	Principal component transformation
pLSA	Probabilistic Latent Semantic Analysis
RIFT	Rotation-Invariant Feature Transform
SEFCM	Separate eigen-based fuzzy C-means

SHVS   Slant horizontal vertical search
SIFT   Scale-Invariant Feature Transform
SLDA   Spatial Latent Dirichlet Allocation
SNR   Signal-to-Noise Ratio
SRKDA   Spectral Regression Kernel Discriminant Analysis
SSS   Square spiral search
SURF   Speeded Up Robust Features
SVC   Scalable Video Coding
SVM   Support Vector Machine
TDP   Transformed Dirichlet Process
VCA   Video Content Analysis
VOP   Video object plane

Chapter 1
Image/Video Segmentation: Current Status, Trends, and Challenges

Hongliang Li and King Ngi Ngan

Abstract Segmentation plays an important role in digital media processing, pattern recognition, and computer vision. The task of image/video segmentation emerges in many application areas, such as image interpretation, video analysis and understanding, video summarization and indexing, and digital entertainment. Over the last two decades, the problem of segmenting image/video data has become a fundamental one and had significant impact on both new pattern recognition algorithms and applications.

This chapter has several objectives: (1) to survey the current status of research activities including graph-based, density estimator-based, and temporal-based segmentation algorithms. (2) To discuss recent developments while providing a comprehensive introduction to the fields of image/video segmentation. (3) To identify challenges ahead, and outline perspectives for the years to come.

1.1 Introduction

We often hear the old adage "a picture is worth a thousand words", which means a complex semantic information can be conveyed with just a single still picture. Have you ever wondered when you look at a picture, how do your eyes find the interesting target from the scene and how do your brain understand the scene? How many activities are involved in the scene recognition progress? The possible answer may lie in the semantic content processing, which can provide us with the meaningful cues for the scene understanding [1].

From the content-related services, a semantic object (i.e., meaningful entity including a collection of attributes) can be detected and exploited to provide the user with the flexibility of content-based access and manipulation, such as fast indexing from video databases, advanced editing and composition, and efficient coding of

H. Li (✉)
School of Electronic Engineering, University of Electronic Science and Technology of China, Chengdu, China
e-mail: hlli@uestc.edu.cn

K.N. Ngan and H. Li (eds.), *Video Segmentation and Its Applications*,
DOI 10.1007/978-1-4419-9482-0_1, © Springer Science+Business Media, LLC 2011

regions of interest [2]. In the past several years, there has been rapid growing interest in content-based applications of video data including video retrieval and browsing, video summarization, video event analysis, and video editing. The requirements for efficiently accessing a great amounts of multimedia content are becoming more and more important. However, how to obtain semantic contents successfully from an image/video is still a very challenging task in the computer vision and pattern recognition.

In order to understand the scene content, we need to known what is the basic component for such content. The common answer may be the semantic object, which represents a data item together with its underlying semantic context. It may consist of a flexible set of meta-attributes that explicitly describe the implicit assumptions about the meaning of the data item [4]. Each semantic object should clearly specify the relationship between the object and the real aspects. Therefore, a crucial step before the image understanding is to separate the image/video into several constituent parts.

In general, segmentation can be defined as the process of partitioning data into groups of potential subsets that share similar characteristics. It has become a key technique for semantic content extraction and plays an important role in digital multimedia processing, pattern recognition, and computer vision. The goal of image segmentation is very application oriented, which emerges in many fields. A limited set of applications of image/video segmentation can be presented as follows:

- *Object recognition*, where the segmentation is treated as a key component that groups coherent image areas that are then used to assemble and detect objects [5]. As important recognition tasks, feature extraction and model matching rely heavily on the quality of the image segmentation process. When an image is segmented into several homogeneous intensity regions, each region can be used as features for deriving the category model since they are rich descriptors, usually stable to small illumination and viewpoint changes [6].
- *Video monitoring*, where an object can be divided into pieces to improve tracking robustness to occlusion by tracking the evolution of the moving objects along the time axis [7]. The segmented mask allows to predict and identify an intruder or of an anomalous situation, and help to reveal their behaviors and make quick decision when "alerts" should be posted to security unit.
- *Video indexing*, which performs over segments of the media using the annotations associated with the segments [8, 9]. An ordered list of segments associated with the query object will be returned to user, which has been applied to the content classification, representation, or understanding.
- *Data compression*, which allows suitable coding algorithm to manipulate each object independently resulting in subjective quality improvement. Segmentation is used to partition each frame of a video sequence into semantically meaningful objects with arbitrary shape. More coding bits can be assigned to these object regions [10], which can reduce visual artifacts after the low-bit rate coding.
- *Computer vision*, where segmented objects from the input 2-D images or video sequences can be used to construct the 3-D scene. For example, stereo for image-based rendering was proposed based on image oversegmentation. Since

entire segments are matched instead of single pixels, the initial match values are more robust to image noise and intensity bias [11].

- *Videophone applications*, which achieve high perceptual quality by coding the areas of interest with better quality. Fewer bits can be allocated for encoding the background by using the higher quantization level. The motivation of this application is that the foreground region is the most important for the viewer [12].
- *Digital entertainment*, such as video matting and video tooning, which employs the segmented objects to generate fantastic effect, or puts them into a virtual scene or game.

There are other possible applications, such as medical diagnosis, tele-education, industrial inspection, environmental monitoring, or the association of metadata with the segmented objects, etc.

Segmentation has become an efficient way to bridge the primary image data and semantic content in image/video processing. In order to satisfy the future content-based multimedia application, more and more researchers seek for efficient ways to segment arbitrary object from multimedia data over the past decade. There are many methods that addressed the segmentation problem, which can be categorized with respect to various criteria:

(1) *Data-based mode*: Based on the original data types, segmentation can be classified into image (e.g., nature, medical, or remote sensing images, etc.), video, audio, and text segmentations, which can be applied to different scenarios. For example, we can use the text segmentation to extract captions displayed in movies, or partition a document into interesting parts. In this chapter, we only concentrate on the image and video segmentation.

(2) *Interaction-based mode*: Two main categories can be classified, namely supervised and unsupervised modes. Supervised methods require user intervention for segmentation, which allow users to easily indicate the foreground across space and time. These methods can provide the better performance than automatic ways because the prior knowledge of the object can be obtained by selecting training data on the images. Unsupervised methods mean that there is no contextual knowledge assumption regarding to the object being segmented. Object segmentation is performed in fully automatic manner, which has become the key technique in a large number of real-time application areas such as video monitoring and surveillance.

(3) *Feature-based mode*: Feature extraction plays important role in image/video segmentation. According to the selection of the feature space, segmentation can be divided into color, texture, intensity, shape, or motion based segmentation method. These features are usually applied to evaluate the region property. For example, for color segmentation, the grouping decision relies on the color distance between neighboring pixels. For the motion segmentation, the main problem is to find independently moving objects in a video in terms of the motion cue.

(4) *Inference-based mode*: Segmentation can be formulated as two message passing modes, namely bottom-up and top-down segmentations. The first performs segmentation on the basis of low-level visual features (e.g., color, texture, intensity,

etc.) rather than high-level knowledge about the object of interest. The resulting segmentation is usually implemented in unsupervised manner. The second method usually requires a database of human-annotated images to learn a prior distribution, which helps to make high-level recognition by incorporating low-level grouping results.

(5) *Space-based mode*: Based on the view of space relation, we can classify segmentation into spatial or temporal methods. The first method focuses on the partition according to the spatial relations among pixels, while the second aims to divide a sequence of frames into several segments along the temporal axis. For example, we can use scene analysis techniques such as video cut, fade, wide, zoom, etc. to perform the scene segmentation so as to group those frames with similar content.

(6) *Class-based mode*: Many segmentation methods are proposed to extract specific objects (e.g., face, human, car, or building) from input images/videos. Since the object is known in advance, the prior information for this object can be used to improve the segmentation results. For example, for the face segmentation, the skin color distribution observed from samples is very helpful for the face region detection, which allows to access the face efficiently.

(7) *Semantic-specific mode*: Unlike the non-semantic segmentation that extracts some uniform and homogeneous segments with respect to texture or color features, semantic segmentation can be defined as a process that typically divides an image into meaningful segments associated with some semantics.

Notice that the existing segmentation methods can be classified into certain categories based on above analysis. Of course, there is no distinct boundary to distinguish different segmentation modes, which means that one can develop a segmentation method by combining different modes. For example, unsupervised over segmentation is usually employed as an important step for the top-down segmentation method that groups those segments into a semantic object.

Because image segmentation is application oriented, it is very difficult to measure a given segmentation quality based on an uniform criteria. This means that "what is a good segmentation?" and "how do we distinguish good segmentations from bad segmentations?" highly depend on the application scenarios. Therefore, many researchers answer the above questions by making some assumptions for the goodness of the segmentation, such as the principle of good continuation states that a good segmentation should have [13].

The goal of this chapter is to review theoretically and practically different methods for image/video segmentation. To achieve this goal, we focus our attention to the task of image/video segmentation only. It may be helpful to the reader to know that there have been many other articles that have reviewed the image segmentation from a variety of perspectives in last decade, such as reviews of image segmentation techniques [14, 15], a survey of ultrasound image segmentation [16], an overview of video segmentation [17]. In this chapter, we not only consider the existing methods that are classics or milestones in the field, but the trends and challenges, which may promote future research work.

This chapter is organized as follows. Section 1.2 reviews the existing algorithms for image/video segmentation. Emerging methods are discussed in Sect. 1.3 to show the trends in image/video segmentation. Finally, Sect. 1.4 summarizes main challenges in the research on segmentation methods and offers the outlook for the future.

1.2 The State-of-The-Art Segmentation Methods

The history of image segmentation (i.e., spatial domain) goes back to the nineteenth century. In existence for over 20 years, image/video segmentation has undergone vast technological progress, which has resulted in a great variety of algorithms. This section focuses on representative technologies and will briefly describe specific examples of where the emergence of the research works began a small revolution in image segmentation. A more detailed description of addressed algorithms can be found in numerous reference articles and books.

1.2.1 Graph-Based Segmentation

1.2.1.1 Graph-Cut Algorithm

In 1989, an interesting work was introduced by Greig [18] that the solution of maximum a posteriori estimation (MAP) for binary images can be exactly computed by graph cut. Unfortunately, this idea did not attract much attention until recent years. The first report on image processing was presented by Boykov and Jolly [19], which applied graph cut to image restoration and interactive image segmentation. Given the subsets of marked object and background pixels, this work used graph cut to find the globally optimal segmentation based on a minimum cut algorithm, which also acts as the foundation work in [20–22].

Given an image, this work created a graph $\mathscr{G} = <\mathscr{V}, \mathscr{E}>$, which can be described by a set of nodes $\mathscr{V}$ (e.g., pixels or regions) and a set of link edges $\mathscr{E}$. In particular, two terminals, i.e., the source terminal s and the sink terminal t, are designed to connect to these nodes. In this graph, all the nodes are connected by two kinds of edges: the bidirectional n-links between two neighboring nodes and the t-links between the nodes and the terminals.

A cut $\mathscr{C}$ is defined as a binary partition of the nodes with two subsets, which can be labeled either as the source terminal (foreground) or sink terminal (background). The goal of graph cut algorithm is to search the best cut that has the globally minimal cost (i.e., the sum of the weights of the edges), which is exactly equal to the maximum flow in the graph [19]. In general, for an image $Z = \{z_i\}$, the cost of a cut can be expressed by an energy function, which can be defined as

$$E(Z) = \sum_{i \in \mathscr{V}} E_1(z_i) + \lambda_1 \sum_{\{i,j\} \in \mathscr{E}} E_2(z_i, z_j), \tag{1.1}$$

where E_1 and E_2 denote the data and smoothness cost functions, respectively.

The first term E_1 is used to set the penalties for assigning each pixel to foreground or background, which reflects how the pixel is close to them. In [19], this term was defined as negative log-likelihoods of histograms for "object" and "background" intensity distributions. Generally, in the interactive method, two distributions can be estimated from the labeled regions by the user's paints. It means that the prior distributions can be estimated from labeled pixels in a weak supervised manner.

The second term $E_2(z_i, z_j)$ is designed to measure the similarity between two nodes z_i and z_j by setting a penalty for a discontinuity between them. This term is close to zero when the distinct boundary is found for nodes z_i and z_j, which means that a larger probability of a cut appears between the adjacent pixels. It can be evaluated by using the local intensity gradient or other regularization-based criteria. In [19], an *ad-hoc* function was used to set the boundary penalties.

The exact solution of the maximum flow problem can be reached by using the max-flow/min-cut algorithm, which has been discussed in [23] in detail. This algorithm tries to find a new augmenting path, which would saturate at least one edge in the route and increase the flow to approach the maximum. When no new augmenting path can be found, the maximum flow is reached, which corresponds to the minimum cut.

Figure 1.1 shows an example of graph cut-based image segmentation. The original image with user's inputs is given in Fig. 1.1a, where the red and blue strokes represent the background and the object, respectively. The defined graph is shown in Fig. 1.1b, which includes two terminals (i.e., object (*flower*) and background (*leaves*)) except for general nodes. Figure 1.1c shows the segmentation result by the graph-cut algorithm. It can be seen that the object *flower* is segmented successfully from the image.

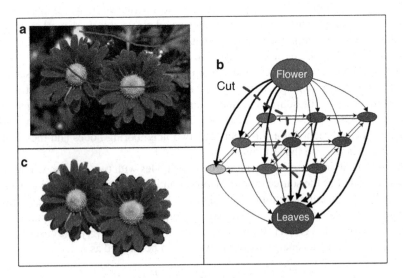

Fig. 1.1 An example of Graph-cut based segmentation. (**a**) Original image *flower* with user's paints. (**b**) The constructed graph with two terminal nodes. (**c**) Segmentation result

1.2.1.2   Random Walks Algorithm

Random walk first appeared in computer vision in the early paper of Wechsler and Kidode for solving the problems of texture discrimination and edge segment detection [25]. In the early works, the random walker algorithm was motivated by placing random walkers at pixels and examining which seeds they first arrive at. However, such a method of computation would be completely impractical. The successful application to the image segmentation was first introduced by Grady's works [26, 27] that apply graph theory to the problems in random walks. To compute the desired probabilities that a walker will first reach the seed with the known label, this work established connections between random walks and the circuit theory (or potential theory) on a graph.

The random walker segmentation is formulated on a graph that is built from an image with a fixed number of vertices and edges. Each edge is assigned a real-valued weight corresponding to the likelihood that a random walker will cross that edge. The detailed algorithm can be summarized as four steps:

(1) *Initialization*: Obtain marked pixels with known labels.
(2) *Mapping*: Map the image to a graph using the typical Gaussian weighting edges.
(3) *Optimization*: Compute the probabilities of unlabeled nodes arriving to each marked label by solving the Dirichlet problem.
(4) *Segmentation*: Obtain a final segmentation by assigning to each node with the class label corresponding to the maximum potential.

Figure 1.2 shows an example of random walker-based image segmentation. The original image with user's inputs is given in Fig. 1.2a, which indicate the background

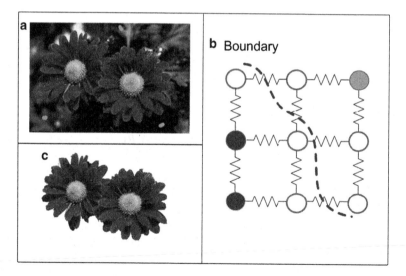

Fig. 1.2   An example of random walker-based segmentation. (**a**) Original image *flower* with user's paints. (**b**) The constructed graph. (**c**) Segmentation result

and the object, respectively. The graph is described in Fig. 1.2b, where the solid and circle nodes denote the known and unknown labels, respectively. Figure 1.1c shows the segmentation result by the random walks algorithm. It can seen that the object *flower* is segmented from the original image.

1.2.1.3 Spectral Clustering Algorithm

Spectral clustering has become one of the most popular modern clustering algorithms, which can be solved efficiently by standard linear algebra methods. Compared to the traditional algorithms (e.g., *k*-means), spectral clustering is very simple to implement and has many fundamental advantages that outperform these approaches [28].

Similar to the graph cut-based method, spectral clustering is also based on weighted graph partitioning ideas, which can be represented in form of a similarity graph $G = (V, E)$ with a set of nodes $V = \{v_1, ..., v_2\}$ and a set of edges E. Each vertex v_i in this graph represents a data point, which connects with its adjacent node by an edge. The main tools for spectral clustering are graph Laplacian matrices, which were studied based on the spectral graph theory. The most common spectral clustering algorithms may include unnormalized spectral clustering, normalized spectral clustering by [29] and [30].

A classic spectral clustering-based segmentation method was first introduced by Shi and Malik in [29] using a normalized cut criteria. Assume a graph G can be partitioned into two disjoint sets A and B by finding a minimum cut. Here a cut is defined as

$$cut(A, B) = \sum_{i \in A, j \in B} w(i, j). \tag{1.2}$$

Generally, it is relatively easy to minimize (1.2) by simply separating data into two parts. However, in practice it often does not lead to satisfactory partitions. To avoid such weird partition, Shi redefined the cut function, by adding some constraints such as the subsets' volume, to normalize the cut cost (i.e., Ncut). The definition can be written by

$$Ncut(A, B) = \frac{cut(A, B)}{vol(A)} + \frac{cut(A, B)}{vol(B)}, \tag{1.3}$$

where $vol(A) = \sum_{i \in A, j \in V} w(i, j)$ denotes the volume of subset A, which represents the relation between nodes in A and nodes in the whole graph. The minimization of (1.3) can be approximated by solving an eigenvalue system [29].

Based on the Ncut algorithm, an image can be segmented into multiple regions by using the following steps:

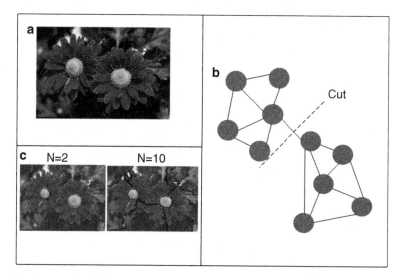

Fig. 1.3 An example of Ncut-based segmentation. (**a**) Original image *flower*. (**b**) The constructed graph. (**c**) Segmentation results with $N = 2$ and $N = 10$

(1) Map an image into a weighted graph $G = (V, E)$ with the nodes corresponding to pixels and weight on the edges setting by the affinity of pairwise pixels.
(2) Construct affinity matrix W and degree matrix D.
(3) Solve the generalized eigenvalue system with the second smallest eigenvector.
(4) Use the eigenvector to partition the graph.
(5) After the stability examination, recursively repartition the segmented parts if necessary.

Figure 1.3 shows an example of Ncut-based image segmentation. The original image is shown in Fig. 1.3a. Since this method is performed in unsupervised manner, there is no user's input for the known labels. The graph is described in Fig. 1.3b, where the dashed line denotes a cut to separate this graph into two parts. Figure 1.3c shows the segmentation results when $N = 2$ and $N = 10$. It can be seen that the original image is segmented into many regions.

1.2.1.4   Efficient Graph Segmentation Algorithm

Another graph-based segmentation algorithm can be found in [24], which computes image segmentation based on pairwise region comparison in unsupervised manner. The boundary between two regions is measured by a predicate based on a graph-based representation of the image. It can be computed using the minimum weight edge between two regions. Although this algorithm makes greedy decisions, it runs in time nearly linear to the number of graph edges (e.g., $O(m \log m)$ time for m graph edges) and is also fast in practice. Different segmentation results can be achieved by setting various parameters including sigma, which is used to smooth the input image and the threshold θ. Some experimental results and the source code can be referred to http://people.cs.uchicago.edu/pff/segment/.

1.2.2 Nonparametric Clustering-Based Segmentation

Mean shift analysis is a nonparametric, iterative procedure introduced by Fukunaga [31] for seeking the mode of a density function represented by local samples, which was generalized by Cheng for the image analysis [33]. More specifically, mean shift estimates the local density gradient of similar pixels via finding the peaks in the local density. It is proved that mean shift procedure is a quadratic bound maximization both for stationary and evolving sample sets [32]. Comaniciu and Meer extended this algorithm to the color image segmentation application [34].

Given n data points x_i in d-dimensional space. The general multivariate kernel density estimator with kernel $K(x)$ is defined as

$$\hat{f} = \frac{1}{n} \sum_{i=1}^{n} K_H(x - x_i). \tag{1.4}$$

For the radially symmetric kernel with the identity matrix $H = h^2 I$, (1.4) can be rewritten by

$$\hat{f} = \frac{1}{nh^d} \sum_{i=1}^{n} K\left(\frac{x - x_i}{h}\right). \tag{1.5}$$

By taking the gradient of (1.5) and employing some algebra, a mean shift vector can be obtained by

$$\mathbf{m}(x) = C\frac{\nabla \hat{f}(x)}{\hat{f}(x)}, \tag{1.6}$$

where C is a positive constant and

$$\mathbf{m}(x) = \frac{\sum_{i=1}^{n} x_i g(\|\frac{x-x_i}{h}\|^2)}{\sum_{i=1}^{n} g(\|\frac{x-x_i}{h}\|^2)} - x. \tag{1.7}$$

Note that the function $g(x)$ is the derivative of the kernel profile $k(x)$, i.e., $g(x) = -k'(x)$.

In general, the kernel $K(x)$ is usually broken into the product of two different radially symmetric kernels, namely the spatial domain and the color range. For a still image, the mean shift segmentation algorithm [35] can be described as following steps:

(1) Given an image, perform the mean shift filtering procedure until convergence.
(2) Grouping together all points that are closer than spatial and range kernel bandwidths.

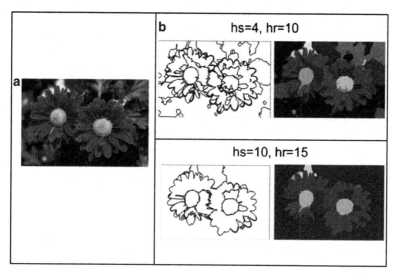

Fig. 1.4 An example of mean shift segmentation. (**a**) Original image *flower*. (**b**) Segmentation result with different parameters

(3) For each group, assign a label.
(4) Eliminate those regions with less pixels.

Figure 1.4 shows an example of mean shift image segmentation. The original image is shown in Fig. 1.4a. Figure 1.3b shows the segmentation results with different kernel bandwidths. The spatial and range bandwidths are set to $h_s = 4, h_r = 10$ and $h_s = 10, h_r = 15$, respectively. We can see that with the increase of kernel bandwidths, more pixels are grouped together, which results in large regions partition. This method is also implemented in an unsupervised manner.

1.2.3   Motion-Based Segmentation

In general, image segmentation algorithms mentioned above can be regarded as spatial-based video segmentation. One can simply perform video segmentation frame by frame using spatial segmentation methods. However, this will result in low efficiency of video segmentation because high correlation between adjacent frames in the temporal axis is neglected.

Temporal segmentation is usually based on change detection followed by motion analysis. The change detection masks can be defined as the absolute difference between two consecutive frames, which are the most common forms of motion information incorporated into the segmentation process. This algorithm employs intensity changes produced by the motion of moving object to identify the position and boundary of objects in time and space. Unlike the spatial segmentation approaches, higher efficiency can be achieved because of small number of operations for the segmented moving region instead of the whole image for every frame [3].

As an important low-level feature, motion can provide the otherwise missing semantic information in cases where uniform motion is expected. In order to extract the moving objects, motion estimation methods are needed especially when change detection masks have been shown to be ineffective. An interesting work on moving object segmentation can be referred to [2]. The core of this algorithm is an object tracker that matches a two-dimensional (2D) binary model of the object against subsequent frames using the Hausdorff distance. To achieve this goal, the first step is to detect a dominant global motion that can be assigned to the background based on the six-parameter affine transformation. An object tracker based on Hausdorff distance is then established to measure the temporal correspondence of objects and enhance the robustness to noise and changes in shape in the video sequence.

1.3 Technological Trends for Image/Video Segmentation

Most past research activities on video segmentation have relied on two principles of spatial (i.e., image) and temporal segmentation. If we treat the motion cue as one of the low level features such as intensity, color, and texture, many image segmentation algorithms can be easily extended to video segmentation. For example, to segment a moving object out from a video clip, a 3D graph cut was presented to partition watershed presegmentation regions into foreground and background while preserving temporal coherence. For each frame, the segmentation in each tracked window is refined using a 2D graph cut based on a local color model [36]. In this section, we will address the following trends for segmentation algorithm especially for the spatial domain segmentation.

1.3.1 Towards 'Good' Segmentation

An emerging trend is to answer the question "What is a good partition for an image?" An interesting work in the current literature is to group pixels into "superpixels", which are local, coherent, and which preserve most of the structure necessary for segmentation at the scale of interest [13, 37]. To generate the superpixel map, the Ncut segmentation algorithm is used by incorporating the contour and texture cues. To find the "good" segmentation, the gestalt grouping cues, such as contour, texture, brightness, and good continuation are combined in a principled way. A linear classifier is trained to combine these features.

An example of superpixel segmentation is shown in Fig. 1.5, which has the number of superpixels 200. The original image flower is shown in Fig. 1.5a, which has the superpixel map given in Fig. 1.5b. A result of segmentation can be found in Fig. 1.5c, which shows that distinct improvement can be achieved with respect to those classic methods.

a   b   c

Fig. 1.5 An example of superpixel segmentation. (**a**) Original image *flower*. (**b**) Superpixel map with the number of 200. (**c**) Segmentation result based on the superpixel map

In addition, there are many approaches that are proposed recently to partition an image into meaningful regions. For example, an unsupervised segmentation algorithm for color image was proposed in [38], which utilizes the gradient information in CIE L*a*b* color space. The initial regions are first created by grouping those nonedge pixels. The regions with similar color and texture are consequently merged to obtain a final segmentation map. Using semantics information, an iterative region growing method was proposed in [39], which is characterized by edge penalty functions within Markov random field context model and region growing technique. This method allows various region features to be incorporated in the segmentation process. In order to allow scene understanding, a region-based model that combines appearance and scene geometry was proposed to partition a scene into semantically meaningful regions [40]. This model is defined in terms of a unified energy function over scene appearance and structure. In addition, an interesting segmentation technique was proposed to partition multivariate mixed data from a lossy data coding/compression viewpoint [41]. This work aims to search the optimal segmentation that minimizes the overall coding length of the segmented data based on the concepts in lossy data compression and rate-distortion theory. From above analysis, we can see that the goal of these works is to achieve 'good' image segmentation in terms of defined decision strategy.

1.3.2   Towards Machine Learning Based Segmentation

Another emerging trend in image/video segmentation is the learning based segmentation, which seeks good segmentation for understanding images and their semantic contents. These methods learn the optimal clustering algorithms from unsegmented, cluttered images using a probabilistic model incorporating both shape model and bottom-up cues (e.g., color, texture, or edge).

Generative models, as an important probabilistic graphical model, are usually applied to represent the process by which images of objects are created. Unlike the grid graph, in this model, each node represents a random variable, and the links express probabilistic relationships between these variables. A typical probabilistic graphical model for a collection of exchangeable discrete data is Latent

Dirichlet Allocation (LDA), which is mainly used to model text corpora based on the bag-of-words assumption. LDA is a text model which was first introduced by Blei [42] to cluster co-occurring words into topics with semantic meanings. Since it enable efficient processing of large collections while preserving the essential statistical relationships, LDA was not only used for text classification and summarization, but also widely used to discover object categories from a collection of images [43].

Some entities are designed to describe the LDA model including "words", "documents" and "corpora". Notice that a document is a sequence of certain words, which are the basic units of discrete data. A collection of certain documents (e.g., M) corresponds to a corpus. The basic idea of LDA is that documents are represented as random mixtures over latent topics, where each topic z is characterized by a distribution over words w [42]. To borrow this algorithm from text literature, many researchers extended LDA model to solve the computer vision problems by mapping the quantized local descriptors (e.g., SIFT descriptors [44]) to "visual words". Each cluster centers after k-means clustering can be regarded as a visual word, which is used to represent a document (e.g., an image) as a histogram of visual words, namely the bag of words. Based on LDA graphic model shown in Fig. 1.6, a generative process for each document in a corpus can be obtained by defining certain distributions, such as $\theta \sim Dir(\alpha)$, $z_n \sim Multinomial(\theta)$. The details of LDA algorithm can be referred to [42]. Given the training data, the LDA model is used to maximize the marginal distribution $p(w|\alpha,\beta)$ via Gibbs sampler.

Since the traditional LDA model only considers the document as a bag of words, spatial relationships among adjacent words are ignored, which results in low accuracy of the recognition tasks. Thus, many researchers considered improving the performance by incorporating the spatial relations into the LDA model. For example, Cao and Fei-Fei introduced a spatially coherent latent topic model (Spatial-LTM) that can improve the traditional bag of words representation of texts and images [45]. In this model, an image is first partitioned into regions, which are described by appearance feature and a set of visual words. Each region is treated as a document. The labels of regions denote the latent topic. The Spatial-LTM model is estimated by the variational message passing algorithm, which can simultaneously segment and classify objects. The similar extension of LDA model can also be found in the Spatial Latent Dirichlet Allocation model [46], which encodes spatial structure among visual words. It clusters visual words that are close in space into one topic.

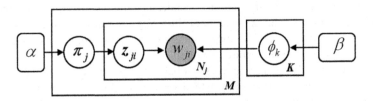

Fig. 1.6 LDA graphical model

1.3.3   Towards Perceptual Model-Based Segmentation

As an important perceptual model, visual attention is an effective simulation of human visual characteristic, which allows us to find relevant information quickly and efficiently. When people gaze at a picture, the human perceptual system always pays more attention to the meaningful objects among the image. We usually call these objects as salient objects. For example, given an image in Fig. 1.7a, the animal (e.g., sheep) should be regarded as the attention object that corresponds to the white region in Fig. 1.7b.

Unlike the traditional methods, attention-based segmentation aims to segment the meaningful physical entities which are more likely to attract users' attention than other objects in the image/video. Figure 1.8 describes a framework of video attention object segmentation, which mainly consists of three steps, i.e., saliency map generation, region segmentation, and object tracking. Given a sequence of images, the first step for this model is to extract the saliency map based on the attentive features that can be modeled by the low-level cues. Apart from spatial features, such as intensity, color, and texture, the temporal feature (i.e., motion) is also important for developing video attention model. The goal of the second step is to group salient pixels by using a clustering method. Finally, object tracking should be used to update object mask for the consequent frames.

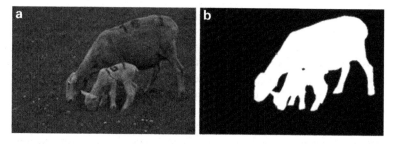

Fig. 1.7 An example of visual attention model. (**a**) Original image *sheep*. (**b**) Ground truth mask of the salient object

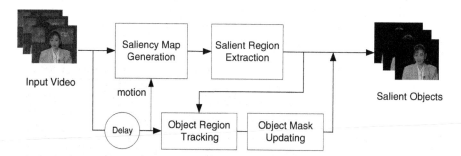

Fig. 1.8 Framework of video attention objects segmentation

To segment attention object successfully, it is important to measure the saliency from the images/video accurately. The well-known visual attention model is called Itti model, which was developed for rapid scene analysis by combining multiscale image features into a single topographical saliency map [47]. A dynamical neural network was used to select attended locations from the saliency map. This work presented a conceptually simple computational model for saliency-driven focal visual attention, which also included some basic concepts for producing some of the performance of primate visual systems, such as center-surround operation and multiscale saliency model. This model was successfully applied to object extraction from color images [48], which formulated the attention objects as a Markov random field by integrating computational visual attention mechanisms with attention object growing techniques. In order to extract visual attention effectively, a lot of methods have been presented recently to deal with salient points detection, such as frequency tuned saliency (FTA) [49], spectral residual saliency [50], site entropy rate [51], and context-aware saliency [52].

In addition, based on the visual attention idea, several object attention models were successfully constructed to extract the object of interest in videos, such as the facial saliency model [53] and focused saliency model [54]. Unlike the general saliency model, object attention model is designed based on the prior knowledge or the training procedure. For example, the first model given in [53] is proposed to segment human face from the head-and-shoulder type video based on a facial saliency map, which is defined as:

$$S(x,y) = P_1(x,y) \cdot P_2(x,y) \cdot P_3(x,y), \tag{1.8}$$

where P_1, P_2, and P_3 denote the "conspicuity maps" corresponding to the chrominance, position, and luminance components, respectively. Each component utilizes the knowledge of human face, such as the skin color that can be detected by the presence of a certain range of chrominance values with narrow and consistent distribution in the YCbCr color space. An example of facial saliency map is shown in Fig. 1.9, where high saliency values usually correspond to the face regions.

In order to highlight the primary objects in an image, the attention object is usually shown in sharp focus, whereas background objects are typically blurred being out-of-focus. The second saliency model [54] was proposed to extract focused objects automatically based on the matting model, which mainly consists of three steps. The first step is to generate a re-blurred version of the input video image by a point-spread function in the proposed method. The focused saliency map of

Fig. 1.9 An example of facial saliency map. *Left*: Original image *claire*. *Right*: The facial saliency map

Fig. 1.10   An example of focus saliency map. *Left*: Original image *gandalf*. *Right*: The focus saliency map

Fig. 1.11   An example of image pairs for co-segmentation. *Left*: Image pair *amira*. *Right*: Image pair *stone*

an image can be computed from the difference between the original and the blurred images. Most of the energy in the saliency map corresponds to the focused object, whilst a large amount of the energy of the defocused region is removed efficiently. In the second stage, bilateral and morphological filtering are employed to smooth and accentuate the salient regions. The third stage involves adaptive error control matting scheme to extract the boundaries of the focused objects. An example of focus saliency map is shown in Fig. 1.10, where most human regions have higher saliency values than background.

1.3.4   Towards Object Driven Segmentation

A typical work of object driven segmentation is called "co-segmentation", which aims to segment the similar object from images. The main idea of co-segmentation is to discover common objects from image pairs based on the assumption that each image contains the foregrounds with similar color, texture, or shape. The first row of Fig. 1.11 shows two image pairs with similar objects in the foreground but different backgrounds, which have been used in many works [55–58]. The second row of Fig. 1.11 shows the ground truth masks for image pairs. Many approaches have been proposed to address the co-segmentation problem in terms of different optimization techniques, such as the L1 norm model [55], L2 norm model [56], and the "reward" model [57].

Generally, the problem of co-segmentation can be formulated as an energy optimization, which can be defined as:

$$x = \arg\min_{x} \underbrace{E_1(x)}_{\text{Intra term}} + \underbrace{E_2(f_1, f_2,)}_{\text{Inter term}}, \qquad (1.9)$$

where x denotes the label set with the value $\{0, 1\}$. f_1 and f_2 are the description for the foreground or background respectively. E_1 is defined as intra penalty within each image, which can be expressed by the MRF including the unary term and pairwise term. The constraint between images is imposed by the second term E_2, which makes the foreground of each images similar with each other. However, the optimization of energy function becomes an NP hard problem. In order to overcome these problems, different optimization methods have been proposed for solving this problem, such as trust region graph cut [55], quadratic pseudo boolean optimization [56], graph cut [57], and dual decomposition [59]. A brief review of co-segmentation can be referred to [59].

Another type of object driven segmentation is the class-specific segmentation, which is to extract the object of interest from the given images/video. An example of such works can be found in [62], which segments human faces automatically. This method proposed an effective segmentation system for cutting human faces out from video sequences in realtime, which consists of three stages. First, a learning based face detector is developed to rapidly identify human faces. To speed up the detection process, a face rejection cascade is constructed to remove most of negative samples while retaining all the face samples. A coarse-to-fine segmentation approach is then used to extract the faces based on a min-cut optimization. Finally, in order to refine the object boundary, this method employed a matting algorithm to estimate the alpha-matte based on an adaptive trimap generation method.

As a highly nonrigid object, human face holds a high degree of variability in size, shape, color, and texture. This method developed a fast face detector shown in Fig. 1.12, which consists of skin color filtering, rejector cascade, and cascades of boosted face classifier. The filter is used to clean up the non-skin regions in the color image during face detection. The rejector is designed to remove most of the non-face candidates while allowing high accuracy for face detection. The promising face-like locations will be examined in the final boosted face classifier. Note that the real-time segmentation system for human face can be easily extended to other applications. For example, if the coarse segmentation is performed on the appropriately defined body region, this work can be extended to solve more challenging "head-shoulder segmentation" problem.

In addition, an early work of class-based segmentation method has been discussed in [60], which aims to capture the common characteristics from a stored

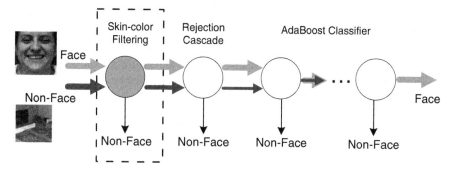

Fig. 1.12 Face detection cascade in [62]

representation of the shape of objects within a general class (such as horse images), and then use this information to segment novel images. An interesting trend can be observed, which aims to segment a collection of unlabeled images while exploiting automatically discovered appearance patterns shared between them [61].

1.4   Challenges

In this section, we identify serious challenges that remain despite over ten years of progress in image/video segmentation.

The first challenge in image segmentation is the semantic gap, which means how to bridge the semantic gap between low-level features and high-level semantic effectively. As stated in the last section, given a similarity matrix (e.g., intensity based), many algorithms can be used to implement data clustering. But it is still difficult to ensure the clustering result with meaningful partitions. For good segmentation, both spatial and semantic relations are required. The demand is significantly urgent for image analysis and scene understanding. Thus far, a range of techniques have been developed to achieve 'good' segmentation by incorporating more semantic information into the clustering process.

The second challenge is to yield accurate segmentation for images. Although image segmentation is application oriented, it is not necessary to provide accurate segmentation mask for some applications. However, accurate segmentation is particularly interesting in a lot of fields. Many are not willing to accept the coarse segmentation results that may be obtained by the existing approaches. More specifically, a robust and accurate segmentation will lead to the distinct improvement for many applications such as content based video coding (e.g., MPEG-4). However, it is still a challenging task to extract accurate object mask from the image/video because of the variations of brightness, lighting, view, and other complex backgrounds.

Despite the advances in video segmentation, it remains difficult to segment objects of interest in real time, which can be regarded as the third challenge. Unlike the static image segmentation, it seems that temporal features such as motion that can be used to improve the video object segmentation. More challenging tasks will be involved due to the inconsistent motion. To achieve real time segmentation, people usually make the trade-off between the segmentation quality and the efficiency, which will directly reduce the performance of the object segmentation. Apart from the problems raised in the image segmentation, fast and accuracy object tracking and mask updating also need to address sufficiently.

The fourth major challenge in image/video segmentation is the need to develop appropriate validation and evaluation approaches. The first task is the formation of common databases where all algorithms can be compared with each other. Fortunately, a good segmentation Dataset and Benchmark was recommended by Martin [64], which contains a number of static images (i.e., 100 test and 200 train images) with different objects and backgrounds. The second issue is the development of evaluation technique. Most evaluation methods in the current literature are based on the computation of the scores between the ground truth mask and the segmented result. It is reasonable but not sufficient to address the segmentation quality. Some researchers in the field have been trying to address this critical issues. One effort that has been carried out in recent years is the work [63], which presented the PR index to compare the obtained segmentation with multiple ground truth images through soft nonuniform weighting of pixel pairs that accounts for scale variation in human perception.

1.5 Summary

Image/video segmentation will have a major role in intelligent visual signal processing in the decades to come. With the transformation of image analysis from human interactive mode toward an unsupervised mode, segmentation is becoming an essential tool for pattern recognition and computer vision. It will be very useful for bridging the semantic gap between the low level feature and the semantic concepts. Segmentation is a fundamental research topic that provides unique opportunities for content based coding and media analysis. However, the challenges listed in the last section should be adequately addressed so as to continue to be a key technology in pattern recognition.

References

1. Ngan King N., Li H.: Semantic Object Segmentation. IEEE Communications Society Multimedia Communications Technical Committee E-Letter, 4(6), 6–8 (2009)
2. Meier T., Ngan K. N.: Automatic segmentation of moving objects for video objects plane generation. IEEE Trans. Circuits and Systems for Video Technology, 8(5), 525–538 (1998)
3. Li H., Ngan King N.: Automatic video segmentation and tracking for content-based applications, IEEE Communications Magazine, 45(1), 27–33 (2007)

4. Bornhovd C., BuchmannA. P.: A Prototype for metadata-based integration of internet sources, Advanced Information Systems Engineering, Lecture Notes in Computer Science, 1626, 439–445 (2010)
5. Kokkinos I., Maragos P.: Synergy between Object Recognition and Image Segmentation Using the Expectation-Maximization Algorithm, IEEE Trans. Pattern Analysis and Machine Intelligence, 31(8), 1486–1501 (2009)
6. Todorovic S., Ahuja N.: Unsupervised category modeling, recognition, and segmentation in images, IEEE Trans. Pattern Analysis and Machine Intelligence, 30(12), 2158– 2174 (2008)
7. Gentile, C., Camps, O., Sznaier, M.: Segmentation for robust tracking in the presence of severe occlusion. IEEE Trans. Image Processing, 13(2), 166–178 (2004)
8. Mezaris V., Kompatsiaris I., Boulgouris N. V., Strintzis M. G.: Real-time compressed-domain spatiotemporal segmentation and ontologies for video indexing and retrieval, IEEE Trans. Circuits and Systems for Video Technology, 14(5), 606–621 (2004)
9. Ko, B., Byun, H.: FRIP: a region-based image retrieval tool using automatic image segmentation and stepwise Boolean AND matching, IEEE Trans. Multimedia, 7(1), 105–113 (2005)
10. Meier T., Ngan K.N.: Video segmentation for content-based coding, IEEE Transactions on Circuits and Systems for Video Technology, 9(8), 1190–1203 (1999)
11. Zitnick C. L., Kang S. B.: Stereo for Image-based rendering using image over-segmentation, International Journal of Computer Vision, 75(1), 49–65 (2007)
12. Chai D., Ngan K.N.: Face segmentation using skin color map in videophone applications, IEEE Transactions on Circuits and Systems for Video Technology, 9(4), 551–564 (1999)
13. Ren X., Malik J.: Learning a classification model for segmentation, Intl Conf. Computer Vision (ICCV), vol. 1, 10–17 (2003)
14. Nikhil R. P., Sankar K. P.: A review on image segmentation techniques, Pattern Recognition, 26(9), 1277–1294 (1993)
15. Cremers D., Rousson M., Deriche R.: A Review of statistical approaches to level set segmentation: integrating color, texture, motion and shape, International Journal of Computer Vision, 72(2), 195–215, (2007)
16. Noble J.A., Boukerroui, D.: Ultrasound image segmentation: a survey, IEEE Trans. Medical Imaging, 25(8), 987–1010 (2006)
17. Koprinska I., Carrato S.: Temporal video segmentation: A survey, Signal processing: Image communication, 16(5), 477–500 (2001)
18. Greig D. M., Porteous B.T., Seheult A.H.: Exact maximum a posteriori estimation for binary images, Journal of the Royal Statistical Society Series B (Methodological), 271–279 (1989)
19. Boykov Y., Jolly M.-P.: "Interactive graph cuts for optimal boundary & region segmentation of objects in n-d images", Proc.ICCV 2001, vol. 1, 105–112 (2001)
20. Rother C., Kolmogorov V., Blake A.: GrabCut: interactive foreground extraction using iterated graph cuts, Proc. SIGGRAPH 2004, (2004)
21. Li Y., Sun J., Tang C.-K., Shum H.-Y.: Lazy snapping, Proc. SIGGRAPH 2004, (2004)
22. Wang J., Bhat P., Colburn R. A., Agrawala M., Cohen M. F.: Interactive video cutout, Proc. SIGGRAPH 2005, (2005)
23. Boykov Y., Kolmogorov V.: An experimental comparison of Min-Cut/Max-Flow algorithms for energy minimization in vision, IEEE Trans. Pattern Analysis and Machine Intelligence, 26(9), 1124–1137 (2004)
24. Felzenszwalb P. F., Huttenlocher D. P.: Efficient Graph-Based Image Segmentation, International Journal of Computer Vision, 59(2), 167–181 (2004)
25. Wechsler H., Kidode M.: A random walk procedure for texture discrimination,IEEE Trans. Pattern Analysis and Machine Intelligence, 1(3), 272–280 (1979)
26. Grady L., Funka-Lea G.: Multi-Label image segmentation for medical applications based on graph-theoretic electrical potentials, Proc. Workshop Computer Vision and Math. Methods in Medical and Biomedical Image Analysis, 230–245 (2004)
27. Grady L.: Random walks for image segmentation, IEEE Trans. Pattern Analysis and Machine Intelligence, 28(11), 1768–1783 (2006)
28. Luxburg U. V.: A tutorial on spectral clustering, Statistics and Computing, 17(4), 395-416 (2007)

29. Shi J., Malik J.: Normalized cuts and image segmentation, IEEE Transactions on Pattern Analysis and Machine Intelligence, 22(8), 888–905 (2000)
30. Ng, A., Jordan, M., Weiss, Y.: (2002). On spectral clustering: analysis and an algorithm, Advances in Neural Information Processing Systems, 14, 849–856, MIT Press.
31. Fukunaga K., Hostetler L.D.: The estimation of the gradient of a density function, with applications in pattern recognition, IEEE Trans. Information Theory, 21, 32–40 (1975)
32. Fashing M., Tomasi C.: Mean shift is a bound optimization, IEEE Transactions on Pattern Analysis and Machine Intelligence, 27(3), 471–474 (2005)
33. Cheng Y.: Mean shift, mode seeking, and clustering, IEEE Trans. Pattern Anal. Machine Intell., 17, 790–799 (1995)
34. Comaniciu D., Meer P. (1997). Robust analysis of feature spaces: color image segmentation, IEEE Conf. Comp. Vis. and Pattern Recogn., Puerto Rico, 750–755.
35. Comaniciu D., Meer P.: Mean shift: a robust analysis of feature spaces, IEEE Transactions on Pattern Analysis and Machine Intelligence, 24(5), 603–619 (2002)
36. Li Y. Sun J., Shum H.-Y.: Video object cut and paste. SIGGRAPH 2005, 24, 595–600 (2005)
37. Mori G., Ren X., Efros A., Malik J.: Recovering Human body configurations: combining segmentation and recognition, IEEE Conf. Comp. Vis. and Pattern Recogn., 2, 326–333 (2004)
38. Ugarriza L. G., Saber E., Vantaram S. R., Amuso V., Shaw M., Bhaskar R.: Automatic image segmentation by dynamic region growth and multiresolution merging, IEEE Trans. Image Processing, 18(10), 2275–2288 (2009)
39. Yu Q., Clausi D. A.: IRGS: image segmentation using edge penalties and region growing, IEEE Trans. Pattern Analysis and Machine Intelligence, 30(12), 2126–2139, (2008)
40. Gould S., Fulton R., Koller D.: Decomposing a scene into geometric and semantically consistent regions, Intl Conf. Computer Vision (ICCV), Kyoto, Japan, 2009.
41. Ma Y., Derksen H., Hong W., Wright J.: Segmentation of multivariate mixed data via lossy coding and compression, IEEE Trans. Pattern Analysis and Machine Intelligence, 29(9), 1546–1562 (2007)
42. Blei D. M., Ng A. Y., Jordan M. I.: Latent dirichlet allocation, Journal of Machine Learning Research, 3:993–1022 (2003)
43. Sivic J., Russell B. C., Efros A. A., Zisserman A., Freeman W. T.: Discovering object categories in image collections. Intl Conf. Computer Vision (ICCV), 2005.
44. Lowe D.: Object recognition from local scale-invariant features. In Proc. ICCV, 1150–1157 (1999)
45. Cao L., Fei-Fei L.: Spatially coherent latent topic model for concurrent object segmentation and classification, In Proc. IEEE Intern. Conf. in Computer Vision (ICCV). 2007.
46. Wang X. Grimson E.: Spatial Latent Dirichlet Allocation, in Proceedings of Neural Information Processing Systems Conference (NIPS), 2007
47. Itti L., Koch C., Niebur E.: "A model of saliency-based visual attention for rapid scene analysis," IEEE Trans. Pattern Anal. Mach. Intell. 20(11), 1254–1259 (1998)
48. Han J., Ngan K. N., Li M., Zhang H.-J.: Unsupervised extraction of visual attention objects in color images, IEEE Trans. Circuits and Systems for Video Technology, 16(1), 141–145 (2006)
49. Achanta R., Hemami S. S., Estrada F. J., Ssstrunk S.: Frequency-tuned salient region detection, IEEE Computer Society Conference on Computer Vision and Pattern Recognition (CVPR), 2009.
50. Hou X., Zhang L.: Saliency detection: a spectral residual approach, IEEE Computer Society Conference on Computer Vision and Pattern Recognition (CVPR), 2009.
51. Wang W., Wang Y., Huang Q., Gao W.: Measuring visual saliency by site entropy rate, IEEE Computer Society Conference on Computer Vision and Pattern Recognition (CVPR), 2010.
52. Goferman S., Zelnik-Manor L.: Context-aware saliency detection, IEEE Computer Society Conference on Computer Vision and Pattern Recognition (CVPR), 2010.
53. Li H., Ngan K. N.: Saliency model based face segmentation in head-and-shoulder video sequences, Journal of Visual Communication and Image Representation, Elsevier Science, 19(5), 320–333 (2008)
54. Li H., Ngan K. N.: Unsupervised Video Segmentation with Low Depth of Field, IEEE Transactions on Circuits and Systems for Video Technology, 17(12), 1742–1751 (2007)

55. Rother C., Minka T., Blake A., and Kolmogorov V.: Cosegmentation of image pairs by histogram matching-incorporating a global constraint into MRFs, IEEE Computer Society Conference on Computer Vision and Pattern Recognition (CVPR), vol. 1, pp. 993–1000, 2006.
56. Mukherjee L., Singh V., and Dyer C. R.: Half-integrality based algorithms for cosegmentation of images, IEEE Computer Society Conference on Computer Vision and Pattern Recognition (CVPR), 2009.
57. Hochbaum D. S., and Singh V., An efficient algorithm for co-segmentation, IEEE International Confernce on Computer Vision (ICCV), 2009.
58. Joulin A., Bach F., and Ponce J.: Discriminative clustering for image co-segmentation, IEEE Computer Society Conference on Computer Vision and Pattern Recognition (CVPR), 2010.
59. Vicente S., Kolmogorov V., and Rother C., Cosegmentation revisited: models and optimization, European Conference on Computer Vision (ECCV), 2010.
60. Borenstein E., Ullman S.: Class-specific, top-down segmentation, European Conference on Computer Vision (ECCV), 2002.
61. Lee Y. J., Grauman K.: Collect-Cut: segmentation with top-down cues discovered in multi-object images, IEEE Computer Society Conference on Computer Vision and Pattern Recognition (CVPR), 2010.
62. Li H., Ngan K. N.: FaceSeg: Automatic Face Segmentation for Real-Time Video, IEEE Transactions on Multimedia, 11(1), 77–88 (2009)
63. Unnikrishnan R., and Hebert M., Measures of similarity, IEEE Workshop on Computer Vision Applications, pp. 394–400, 2005.
64. Martin D., Fowlkes C., Tal D., and Malik J.: A database of human segmented natural images and its application to evaluating segmentation algorithms and measuring ecological statistics, in Proc. Intl Conf. Computer Vision (ICCV), vol. 2, pp. 416–423, 2001.

Chapter 2
Image Segmentation with Eigen-Subspace Projections

Jar-Ferr Yang and Shu-Sheng Hao

Abstract In this chapter, object segmentation algorithms dependent on the characteristics of eigen-structure are proposed. The eigen-subspaces are obtained from eigen-decomposition of the covariance matrix, which is computed from the selected color samples. Hence, the color space can be transformed into the signal subspace and its orthogonal noise subspaces. After statistical analysis of eigen-structure of target color samples, the color eigen-structure segmentation algorithms are then designed to extract the desired objects, which are close to the color samples. The principal component transformation (PCT) techniques, which only use the signal subspace can be treated as a subset of color eigenspace algorithms. The eigenspaces discriminated as signal and noise subspaces related to original color samples should be effectively utilized. The adaptive eigen-subspace segmentation (AESS) algorithm, which can save the computation of eigen-decomposition, is applied to adaptively adjust the eigen-subspaces. Finally, the Eigen-based fuzzy C-means (FCM) clustering algorithm has been proposed to effective segment color object. By jointly consideration of signal and noise subspace projections of desired colors, the separate eigen-based FCM (SEFCM) and coupled eigen-based FCM (CEFCM) are used to achieve effective color object segmentation. With these proposed algorithms, the color objects can be successfully extracted by using eigen-subspace projections.

2.1 Overview

Image segmentation has been treated as a key technology in many smart image and video related applications. To achieve effective coding, for example, the image segmentation is the most important kernel in construction of the MPEG-4 video object plane (VOP) [1, 2]. The four major features, including luminance, motion, color,

J.-F. Yang (✉)
Department of Electrical Engineering, National Cheng Kung University, Tainan, Taiwan
e-mail: jfyang@ee.ncku.edu.tw

K.N. Ngan and H. Li (eds.), *Video Segmentation and Its Applications*,
DOI 10.1007/978-1-4419-9482-0_2, © Springer Science+Business Media, LLC 2011

and depth information, have been used as indexes for segmentation of the scenes. If we can separate the video objects and encode them in the video bitstreams, we can achieve the goals such as content scalability, sprite construction, and depth map estimation [3]. Recently, there are many segmentation researches proposed by using motion [4,5], edge [6,7], shape [8], and textual [9,10] information. In this chapter, we introduce the object segmentation algorithms by only using the characteristics of eigenstructure of the color space. After statistical analysis of eigen-structure of the color samples, the color eigen-structure segmentation algorithms, which consider characteristics of signal and noise subspaces, are suggested. By using color information only, simulations show that the proposed algorithm can successfully detect the desired objects from standard video test sequences. In Sect. 2.2, the object segmentation algorithm based on the color eigen-structure characteristics will be stated. In Sect. 2.3, color object segmentation using adaptive eigen-subspaces will be discussed. In Sect. 2.4, color object segmentation using fuzzy C-means with eigen-subspace projection will be described. Conclusions will be stated in Sect. 2.5.

2.2 An Object Segmentation Algorithm Based on Color Eigen-Structure Characterizations

As mentioned in Sect. 2.1, the principal component transformation (PCT) for image segmentation has been proposed [11–13]. The PCT essentially exhibits a color transformation to the signal subspace only. In this section, we propose a color eigen-structure algorithm to efficiently and effectively retrieve the desired objects. In Sect. 2.2.1, the theory of PCT will be introduced. In Sect. 2.2.2, we further analyze the properties of eigen-subspaces. In Sect. 2.2.3, we adopt the statistical analysis of the eigen-structure to design a color eigen-structure segmentation algorithm. The detailed procedures of the algorithm are also described. In Sect. 2.2.4, simulation results will be shown to verify the above theoretical development.

2.2.1 Principal Component Transformation (PCT)

The principal component transform (PCT) [17–19] is also called discrete Karhunen-Loève (KL) expansion. The KL transformation achieves optimal energy compaction and independent properties, which are commonly used for data compression. For the purpose of color object segmentation, the PCT could help to identify the most likely component. The proposed algorithms can also apply to other color coordinates, for example, YUV or YC_rC_b as well. Without losing the generality, we choose RGB components to form the covariance matrix related to the selected color samples. First, using mouse clicks on the desired object to choose a few desired color samples. The kth sample in the RGB color vector is given by

$$\mathbf{s}_k = [r_k \ g_k \ b_k]^T, \tag{2.1}$$

where r_k, g_k, and b_k are red, green, and blue levels of the kth sample in each color plane. In (2.1), the superscript T denotes the transpose of the argument vector. Given M color samples, we can compute the covariance matrix, $\mathbf{R}_s$ as

$$\mathbf{R}_s = \frac{1}{M} \sum_{k=1}^{M} \mathbf{s}_k \mathbf{s}_k^T . \tag{2.2}$$

Applying the eigen-decomposition procedure on the matrix $\mathbf{R}_s$, we can obtain three eigenvectors $\mathbf{w}_1$, $\mathbf{w}_2$, and $\mathbf{w}_3$. The eigenvectors are corresponding to the eigenvalues λ_1, λ_2, and λ_3, which are arranged in the descending order as

$$\lambda_1 \geq \lambda_2 \geq \lambda_3. \tag{2.3}$$

The covariance matrix $\mathbf{R}_s$ can be expressed by

$$\mathbf{R}_s = \sum_{i=1}^{3} \lambda_i \mathbf{w}_i \mathbf{w}_i^T . \tag{2.4}$$

The first principal component $\mathbf{w}_1$ corresponding to the largest eigenvalue becomes the best representation of the desired data samples. If any unknown color samples possess large projections along $\mathbf{w}_1$, we may treat those samples to have a higher possibility with the same classification in color as the selected samples. In order to obtain more satisfactory results, we should jointly consider $\mathbf{w}_1$, $\mathbf{w}_2$, and $\mathbf{w}_3$ projections. We can divide the projected space into two kinds of subspaces, i.e. signal subspace and noise subspaces. The signal subspace is formed by the eigenvector $\mathbf{w}_1$ associated with the largest eigenvalue λ_1 while the noise subspaces are constructed by the eigenvectors $\mathbf{w}_2$ and $\mathbf{w}_3$ corresponding to λ_2 and λ_3. It is noted that the eigenvectors, $\mathbf{w}_1$, $\mathbf{w}_2$, and $\mathbf{w}_3$ of any covariance matrix are orthonormal vectors. Thus, the signal and noise subspaces are orthogonal with each other.

For most PCT methods [14–16], that used the first principal component for color object extraction, they would face the problem in determination of their thresholds by statistical analyses of eigen-structures [17]. For semiautomatic color object segmentation, the sampled pixels could be obtained from mouse clicks upon the desired color objects. With the computed or prestored eigen-structures, the PCT method can extract the features in some conditions [18–20]. In order to localize the desired object in the image, the adaptive eigen-subspaces method will be used to extract the interesting color object [21]. However, the detection performance of the PCT method will be degraded if the color samples are not properly adopted. To achieve satisfactory segmentation, we should further cooperate with iterative or fuzzy inferences to improve the PCT method [22]. In order to extract meaningful objects in different images, we can collect all desired colors to setup color subspaces for initialization of the fuzzy clustering algorithms.

2.2.2 Color Eigen-Subspaces

Assumed the desired objects exhibit an average color sensation, which previously is expressed as (2.1). It is noted that the derivations can be applied to any other color space. However, we develop the algorithm for the RGB color space only. In order to divide the three-dimension color spaces into noise and signal subspaces, we further assume that the desired objects contain no more than two large-displaced colors in the average sense. In other words, the number of the desired colors is limited to $p = 1$ or 2. As to the texture or the shadow effect of the desired objects, the variation of colors in the desired objects are modeled as independent noises and expressed by

$$\mathbf{g}_k = \mathbf{s}_k + \mathbf{n}_k == \begin{bmatrix} r_k & g_k & b_k \end{bmatrix}^T + \begin{bmatrix} n_{r,k} & n_{g,k} & n_{b,k} \end{bmatrix}^T, \qquad (2.5)$$

where $n_{r,k}$, $n_{g,k}$, and $n_{b,k}$ are the kth sampled color noises, which are assumed to be statistically independent to the desired color vector $\mathbf{s}_k$ and uncorrelated with each other. The covariance matrix $\mathbf{R}_s$ of the sampled color vectors is defined as

$$\mathbf{R}_g = E\left[\mathbf{g}_k\mathbf{g}_k^T\right]. \qquad (2.6)$$

Since the number of average color vectors, p is limited under two, i.e., $p = 1$ or 2, the noise free color covariance matrix can be expressed by p principal components as

$$\mathbf{R}_g = \sum_{i=1}^{p} \lambda_i \mathbf{v}_i \mathbf{v}_i^T, \qquad (2.7)$$

where λ_i represents the ith eigenvalue of $\mathbf{R}_g$ and $\mathbf{v}_i$ denotes its corresponding eigenvector. The span of $\mathbf{s}_i, i = 1, \dots, p$ is equal to the span of $\mathbf{v}_i, i = 1, \dots, p$, which is called the signal subspace. Due to the independent assumption of sample noises, the covariance matrix of the sample noises can be modeled as

$$\mathbf{R}_n = \sigma_n^2 \mathbf{I}. \qquad (2.8)$$

The covariance matrix of sampled color vectors composed of both signal and noise components can be expressed by

$$\mathbf{R}_g = \mathbf{R}_s + \mathbf{R}_n = \sum_{i=1}^{p} \lambda_i \mathbf{v}_i \mathbf{v}_i^T + \sigma_n^2 \mathbf{I} = \sum_{i=1}^{p} (\lambda_i + \sigma_n^2) \mathbf{v}_i \mathbf{v}_i^T + \sum_{i=p+1}^{3} \sigma_n^2 \mathbf{v}_i \mathbf{v}_i^T. \qquad (2.9)$$

It is noted that the random noises in the average sense do not change the direction of original signal subspace but add the noise power (variation) σ_n^2 to the true eigenvalues of $\mathbf{R}_s$. The remaining subspaces in the RGB color coordinate system, which are called the noise subspaces, become the span of $\{\mathbf{v}_i, i = (p+1), \dots, 3\}$. It is obvious that the eigenvectors of a symmetrical matrix are orthogonal to each other. Accordingly, the signal subspace and noise subspace will be orthogonal to each other. For example, if we choose the skin as the desired objects by choosing $p = 1$,

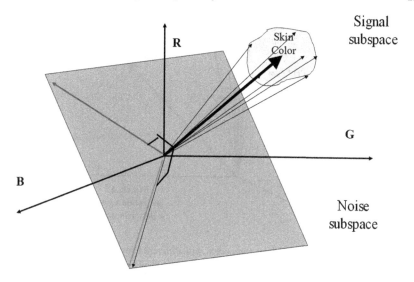

Fig. 2.1 The relationship of signal and noise subspaces

Fig. 2.2 Problems in the first principal component with larger projection (The solid region (dark blue) and dot region (blue) have the same direction)

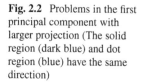

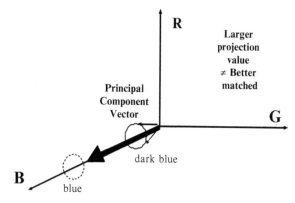

the signal is the span of $\{\mathbf{v}_1\}$ and the noise subspaces become the span of $\{\mathbf{v}_2, \mathbf{v}_3\}$. Figure 2.1 shows the relationship of signal and noise subspaces for the skin objects.

In order to segment the desired object, the most frequent approaches perform the so-called principal color segmentation [23–27]. First, we can obtain the sampled covariance matrix followed by an eigen-decomposition to obtain $\mathbf{v}_1$, the eigenvector, which is corresponding to the largest eigenvalue. With the principal component vector at hand, we then project all the color vectors of image pixels to the principal component vector $\mathbf{v}_1$. Finally, the object segmentation can be achieved by choosing the pixels, which have the largest projections with a proper threshold. The threshold method of the principal value technique is widely adopted for many optimization applications. However, there are two problems that arise in use of the first principal color component for color object segmentation. Figure 2.2 shows the fact that the larger the projection, it implies the better match of color. From the viewpoint of

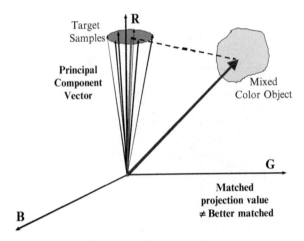

Fig. 2.3 Problem in the first principal component with different hues

color perception, the strong intensity color should be totally different from the low intensity color even if they share the same color space. To classify the dark-blue cloth color, the light-blue sky color, however, will have larger projection. Figure 2.3 shows the case that the mixed colors could have the same color projection although they have different hues. The pure yellow color (R+G) and the pure red color will have totally the same projection if the pure red color object is selected. The highly mixture color such as the high intensity white color, usually has larger projection. Hence, we need exploit the complete eigen-structure to develop a color segmentation algorithm to overcome those deficiencies in using the principal component approaches.

2.2.3 Color Eigen-Subspaces Segmentation

For given sampled color vectors $\mathbf{g}_k$ of the target objects, we can obtain the sampled covariance matrix $\hat{\mathbf{R}}_g$ as (2.6). Through the eigen-analysis procedures, we can obtain the eigenvectors $\hat{\mathbf{v}}_1$, $\hat{\mathbf{v}}_2$, and $\hat{\mathbf{v}}_3$ corresponding to the eigenvalues, $\hat{\lambda}_1$, $\hat{\lambda}_2$, and $\hat{\lambda}_3$, which are arranged in the descending order as

$$\hat{\lambda}_1 \geq \hat{\lambda}_2 \geq \hat{\lambda}_3. \tag{2.10}$$

Since the covariance matrix $\hat{\mathbf{R}}_g$ can be expressed by

$$\hat{\mathbf{R}}_g = \sum_{i=1}^{p} (\hat{\lambda}_i + \hat{\sigma}_n^2)\hat{\mathbf{v}}_i\hat{\mathbf{v}}_i^T + \hat{\sigma}_n^2 \sum_{i=p+1}^{3} \hat{\mathbf{v}}_i\hat{\mathbf{v}}_i^T. \tag{2.11}$$

The least eigenvalue of the sampled covariance matrix can be used as the estimator of the noise power as

$$\hat{\sigma}_n^2 \approx \hat{\lambda}_3, \tag{2.12}$$

and the corresponding eigenvector is estimated by $\mathbf{v}_3 \approx \hat{\mathbf{v}}_3$. The estimated principal eigenvalues are given by

$$\lambda_1 \approx \hat{\lambda}_1 - \hat{\lambda}_3, \tag{2.13}$$

and

$$\lambda_2 \approx \hat{\lambda}_2 - \hat{\lambda}_3. \tag{2.14}$$

While the estimated eigenvectors are $\mathbf{v}_1 \approx \hat{\mathbf{v}}_1$ and $\mathbf{v}_2 \approx \hat{\mathbf{v}}_2$. Before the development of our segmentation algorithm, we should further analyze the statistical properties of the above estimators such that the thresholds of the signal and noise subspaces are reasonably designed.

2.2.3.1   Statistical Analysis of Eigen-Structures

The covariance matrix of the interested spatial samples is given by $\mathbf{g}_k$ as in (2.5). The covariance matrix $\mathbf{R}_g$ and its estimated matrix $\hat{\mathbf{R}}_g$ are respectively defined in (2.9) and (2.11). Based on $\hat{\mathbf{R}}_g$, we can obtain the estimated signal and noise subspaces. To explore their expectations and deviations, we should first analyze the asymptotic statistics for the eigenvalues and eigenvectors of the sampled covariance $\hat{\mathbf{R}}_g$ under the Gaussian process assumption [28]. Based on the perturbation formulation, the first- and second-order moments of $\hat{\lambda}_i$ and $\hat{\mathbf{v}}_i$ can be obtained [20, 29]. The eigenvectors of the signal subspace $\hat{\mathbf{v}}_i$ and its associated eigenvalues $\hat{\lambda}_i$ are asymptotically normal with noise subspaces $\hat{\mathbf{v}}_j$ and $\hat{\lambda}_j$, for $i, j = 1, 2, 3, i \neq j$. According to [20], the expectation value of $\hat{\mathbf{v}}_i$ and the covariance of $\hat{\lambda}_i$ can be expressed by (2.15) and (2.16) as follows:

$$E[\hat{\mathbf{v}}_i] \approx \mathbf{v}_i - \frac{1}{2} \sum_{j=1, j \neq i}^{3} \frac{\lambda_i \lambda_j}{(\lambda_j - \lambda_i)^2 N} \mathbf{v}_i, \tag{2.15}$$

$$\mathbf{cov}(\hat{\lambda}_i, \hat{\lambda}_j) \approx \frac{\delta_{i,j} \lambda_i^2}{N}, \tag{2.16}$$

where N is the number of sampled pixels and $\delta_{i,j}$ is the Kronecter delta. The estimated values of $\hat{\mathbf{v}}_i$ and $\hat{\lambda}_i$ can be expressed by

$$\hat{\lambda}_i = \lambda_i + \xi_i, \tag{2.17}$$

and

$$\hat{\mathbf{v}}_i = \mathbf{v}_i + \mathbf{s}_i, \tag{2.18}$$

where the error terms, $\mathbf{s}_i$ and ξ_i have the following asymptotic properties [20]:

$$\sigma_{\lambda_i \lambda_j}^2 = E[\xi_i \xi_j] \approx \frac{\lambda_i^2}{N} \delta_{i,j}, \tag{2.19}$$

$$E[\mathbf{s}_i] \approx -\frac{\lambda_i}{2N} \sum_{k=1, k \neq i}^{3} \frac{\lambda_k}{(\lambda_i - \lambda_k)^2} \mathbf{v}_i. \tag{2.20}$$

The noise term ξi in (2.17), which has been obtained from (2.19) will be used to set the threshold values on three transformed color spaces in the following section.

2.2.3.2 Object Segmentation Algorithm Based on Color Eigen-Structures

In order to classify the input color vector into the signal and the noise subspace, we can project the color vector on the eigenvectors of the sample covariance matrix as

$$y_{i,k} = \mathbf{v}_i^T \cdot \mathbf{g}_k, \quad \text{for } i = 1, 2, 3. \tag{2.21}$$

Now, we should statistically analyze the projection length of $y_{i,k}$ by taking expectation of the power as

$$E[y_{i,k} y_{i,k}^T] = E\left[\mathbf{v}_i^T \mathbf{g}_k \mathbf{g}_k^T \mathbf{v}_i\right] = \mathbf{v}_i^T \hat{\mathbf{R}}_g \mathbf{v}_i = \hat{\lambda}_i. \tag{2.22}$$

Thus, the average length of the eigenvector projection should become

$$|y_{i,k}| = \left|\mathbf{v}_i^T \cdot \mathbf{g}_k\right| = \sqrt{\hat{\lambda}_i}, \quad \text{for } i = 1, 2, 3. \tag{2.23}$$

For the principal component approach, we can simply detect the color pixel by choosing

$$|y_{1,k}| = \left|\mathbf{v}_1^T \cdot \mathbf{g}_k\right| \geq \sqrt{\hat{\lambda}_1}. \tag{2.24}$$

This is the so-called signal subspace projection. Any pixel color vector, which has large enough projection onto the direction of the principal color vector, will be treated as the object pixel for color segmentation. As Fig. 2.2 shown, the brighter color generally has larger projection. As shown in Fig. 2.3, the mixed color could have the same projection as the target color space. The approach of principal component usually preserves the desired color object, however, erroneously includes brighter color and mixed color pixels.

From the statistics analysis obtained in Sect. 2.2.3.1, we should classify the color space by using both signal and noise subspaces by determining the threshold values in the transformed color spaces. First, we should detect the signal space component more precisely. In order to include 97.5% confidence interval of the principal projection, we propose the signal-subspace detection criterion by modifying (2.24) as

$$\sqrt{\hat{\lambda}_1 + k_{1s}\sigma_{\lambda_1}} \geq |y_{1,k}| \geq \sqrt{\hat{\lambda}_1 - k_{1s}\sigma_{\lambda_1}}, \tag{2.25}$$

where k_{1s} is a constant that equals to 3. The deviation σ_{λ_1} of the first principal eigenvalue is given by

$$\sigma_{\lambda_1}^2 = E[\xi_1 \xi_1] \approx \frac{\lambda_1^2}{N}. \tag{2.26}$$

We relax the lower bound by three deviations to include the possible shadow colors and add the upper bound with three deviations to exclude the unwanted brighter colors. Thus, we can eliminate the incorrect luminance pixels as possible. The pixels, which meet the signal subspace criterion stated in (2.25), could be very possible mixed color pixels.

In order to further exclude the mixed color pixels, we should use the noise space criterion to remove the pixel color pixels from the signal subspace pixels, which satisfy the criterion stated in (2.25). The noise subspace criterion can be discussed in two cases: $p = 1$ and $p = 2$. For $p = 1$, the noise subspace now becomes the span of $\{v_1, v_2\}$. We should perform the noise subspace criterion as

$$|y_{i,k}| = |\mathbf{v}_i^T \cdot \mathbf{g}_k| > \sqrt{\hat{\lambda}_i + k_{\text{in}}\sigma_{\lambda_i}}, \quad \textbf{for } i = 2,3, \tag{2.27}$$

to remove the unwanted pixels, where k_{in} is a constant to specify the confidence interval of the noise. We know that the pixels, whose projections to the noise subspace should be as small as $\sqrt{\hat{\lambda}_i}$ for $i = 2,3$, are matched with the desired color modal. For any other pixels with mixed colors, their color vectors project onto the noise subspace will be larger than $\sqrt{\hat{\lambda}_i}$ for $i = 2,3$. Similarly, we can keep the desired pixels once we find their projections to the noise subspace are beyond the limits of $k_{\text{in}} \cdot \sqrt{\hat{\lambda}_i \pm k_{2n} \cdot \sigma_{\lambda_i}}$ for $i = 2,3$. For $p = 1$, we perform the detection of

$$k_{1n} \cdot \sqrt{\hat{\lambda}_i + k_{2n} \cdot \sigma_{\lambda_2}} \geq |y_{2,k}| = |\mathbf{v}_2^T \cdot \mathbf{g}_k| \geq k_{1n} \cdot \sqrt{\hat{\lambda}_i - k_{2n} \cdot \sigma_{\lambda_2}} \tag{2.28}$$

and

$$k_{1n} \cdot \sqrt{\hat{\lambda}_i + k_{2n} \cdot \sigma_{\lambda_3}} \geq |y_{3,k}| = |\mathbf{v}_3^T \cdot \mathbf{g}_k| \geq k_{1n} \cdot \sqrt{\hat{\lambda}_i - k_{2n} \cdot \sigma_{\lambda_3}} \tag{2.29}$$

to remove the unwanted pixels. We set the constants $k_{1n} = 3$ and $k_{2n} = 3$ in our experiment to achieve the best results. In (2.28) and (2.29), σ_{λ_2} and σ_{λ_3} denote the deviation of the second and third eigenvalues respectively given by

$$\sigma_{\lambda_2}^2 = E[\xi_2 \xi_2] \approx \frac{\lambda_2^2}{N}, \tag{2.30}$$

and

$$\sigma_{\lambda_3}^2 = E[\xi_3 \xi_3] \approx \frac{\lambda_3^2}{N}. \tag{2.31}$$

In summary, we utilize the complete projections of both signal and noise subspaces to detect the desired color pixels. The signal-subspace projection helps to classify the desired pixels while the noise-subspace projection provides the information to eliminate the unmatched pixels. Theoretically, the noise plane related to the smallest eigenvalue can highlight the most wanted object after removing

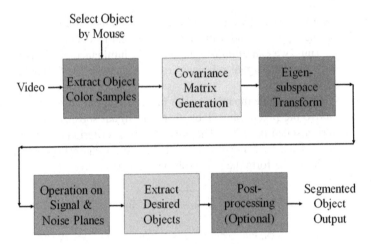

Fig. 2.4 Simulation function block diagram of video object segmentation

noise. After join considering with signal space, the smallest eigenspace is too sensitive to extract the objects that contains too many shadows. These shadows usually are with a similar hue as the desired samples that we should take them as parts of the objects. In order to correct representing the segmented objects, we can take the advantage that the extracted color from the second large noise plane is not so accurate as the smallest one. By inspecting the second noise space related to the second large eigenvalue, we find out that this plane can well describe the silhouette of video object. Hence, the proposed color segmentation using complete eigen-structure should result effective and efficient detection performances in detecting the desired color objects. Figure 2.4 shows the procedures of applying the proposed color eigen-structure segmentation method on some known or instant samples of the desired color object such as skin, hair, or clothes.

2.2.4 Simulation Results

In order to verify the effectiveness of the proposed algorithms, we adopt four standard sequences, Mosaic, Ball, Akiyo, and News, which are shown in Fig. 2.5 in simulations. For each image, we mark with a triangular sign on it to indicate the desired color object we want to extract. Because we have not introduced any spatial and temporal information, all the other similarly colored objects with the same color sensation will also appear in the segmentation results. To clearly exhibit the desired color in the segmented images, we only show the desired color and undesired color objects by bright pixels with gray level = 255 and dark pixels with gray level = 0, respectively. First, we only apply the PCT algorithm to extract the desired color objects. From the signal and noise subspaces, the segmented objects are varied diversely by different threshold setting. By using the threshold determination

Fig. 2.5 Test sequences: (**a**) Mosaic; (**b**) Ball; (**c**) Akiyo; (**d**) News (The desired color objects are marked with triangles)

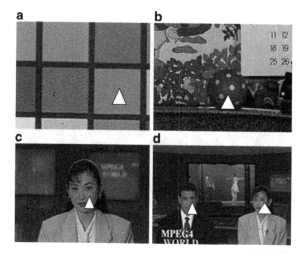

Fig. 2.6 Segmented images obtained by the PCT method using threshold and logical operation: (**a**) Mosaic; (**b**) Ball; (**c**) Akiyo; (**d**) News sequence

suggested in [17] and applying some logical operations, we can obtain four segmented images as shown in Fig. 2.6. Although the main parts of the desired object are extracted, noise cannot be easily de-correlated from signal.

The test sequences are shown in Fig. 2.5 embedded with different characteristics. The scenes of these four sequences are quite different. Akiyo sequence has a blue static TV screen on her background and with some background color similar to Akiyo's face. Carphone sequence has large head movement and fast scene changes outside the car window. Due to fast head movement, the lightness changes occur on his face very quickly. This lightness effect is usually hard to overcome by other

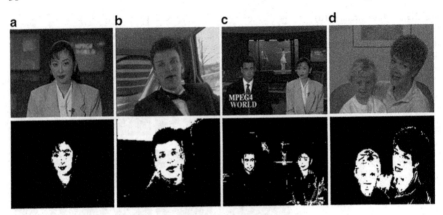

Fig. 2.7 Simulation results of four test sequences: (**a**) Akiyo; (**b**) Carphone; (**c**) News; (**d**) Mother-and-daughter sequences (*up*: original images; *down*: extracted skin color objects)

segmentation methods. The News sequence has two newscasters that occupy two small regions in the scene. In this sequence, the background is more complex with one static blue screen and a large scene-changing TV screen. Actually, the face skin regions are very small in this sequence. The MD sequence has serious shadow effect on mother's clothes and daughter's head.

While simulation, we extract the desired samples only from the first frame of the sequences to obtain the covariance matrix. The skin color extraction results of the first frame from four test sequences are shown in Fig. 2.7. The images on left column show the original images while the right column exhibits the extracted skin color regions. Simulation results show that those skin regions are successfully extracted by using our algorithm.

Inspecting Fig. 2.7, our algorithm can also identify even the small regions such as eyes and mouth, which are with a different color sensation from the skin. In Fig. 2.7b, we also extract the car's roof because it has similar color as the man's face. Applying the temporal redundancy, of course, we can remove the car's roof by using some motion information. In Fig. 2.7d, the extraction results of Mother's and daughter's faces are influenced by the shadows but the main parts of skin are revealed. Figure 2.8 shows the extraction results of clothes and hair objects. Inspecting Fig. 2.8a,b, we can extract the Akiyo's clothes and hair separately according to different sample color. We can also extract the clothes of News and MD as shown in Fig. 2.8c,d. In order to verify the robustness of our algorithm, we take different frames in Akiyo's sequence with the same transformation matrix obtained from the first frame. From Fig. 2.9, we find that our algorithm can mostly extract the skin regions, in which her different expressions can be also observed.

Generally speaking, it is almost unnecessary to perform any post processing method in our algorithm. Inspecting the simulation results, even the small features such as eyes and mouth can be also indicated. If needed, we can also utilize the temporal information to remove the unwanted static scenes. The signal and noise subspaces' thresholds can be defined according to (2.25), (2.28), and (2.29). In

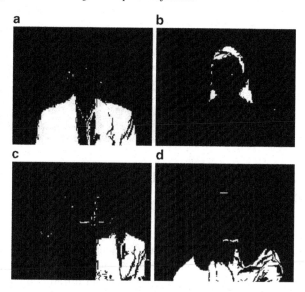

Fig. 2.8 Extracted results of clothes and hair in (**a**) Akiyo (clothe); (**b**) Akiyo (hair); (**c**) News (clothe); (**d**) Mother-and-Daughter (clothe)

Fig. 2.9 Extracted results with same threshold $k_{1n} = 2$, $k_{2n} = 3$, $k_{1s} = 3$ in different frames: (**a**) frame 10; (**b**) frame 30; (**c**) frame 50; (**d**) frame 70

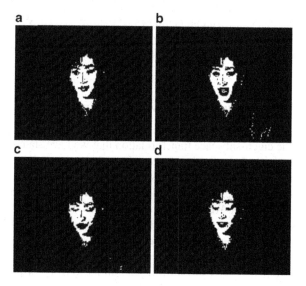

order to verify the detection criterion, we need to explore the different deviations, which will significantly influence the formation of signal subspace applying $k_{1n} = 2$, $k_{2n} = 3$, $k_{1s} = 3$ and $k_{1n} = 1$, $k_{2n} = 3$, $k_{1s} = 3$ as described in (2.25), (2.28), and (2.29) to the Akiyo sequence. Figure 2.10b shows that $k_{1n} = 2$ with $\pm 3\sigma_{\lambda_2}$ deviation achieve better segmentation than that with $k_{1n} = 1$ with $\pm 3\sigma_{\lambda_2}$ as in Fig. 2.10a, where the signal deviation is $\pm 3\sigma_{\lambda_1}$ in both cases.

Fig. 2.10 Extracted results
by different thresholds with
(a) $k_{1n} = 1$, $k_{2n} = 3$, $k_{1s} = 3$;
(b) $k_{1n} = 2$, $k_{2n} = 3$, $k_{1s} = 3$

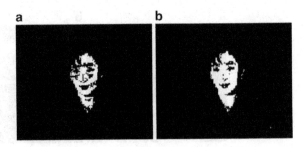

2.3 Color Object Segmentation Using Adaptive Eigen-Subspaces

In the previous section, we have shown that the eigen-subspace is an effective method for separating the signal and noise components. While segmenting the color video objects, several difficulties could be met. If the object is even originally with the same color, it could show distinct properties under different lightening conditions such as shade. In order to solve such problem, an adaptive eigen-subspace segmentation (AESS) algorithm is proposed [21]. The proposed method can estimate and adaptively adjust the eigenvectors under segmentation procedure. Although the object color is changed with different shade, it is still successfully extracted by using this algorithm. Accompaning with the proposed AESS algorithm, three searching algorithms are used to effectively and efficiently locate the possible pixel. Both AESS and the proposed searching algorithms will be discussed in the following sections.

2.3.1 Adaptive Eigenvector Estimation

The eigen-subspace transformation that we have stated in the previous section was applying the same eigenvectors through the simulation. Sometimes, the eigenvectors need to be adaptively adjusted according to different simulation conditions. It is difficult to segment the color object with different shade by just using the same eigenvectors. We introduce a method that can adaptively adjust the eigenvectors in simulation. As stated in (2.1), we can choose an initial RGB color pixel to form a vector s_k. Then, the related covariance matrix $\mathbf{R}_s$ can be obtained using (2.2). With covariance matrix, $\mathbf{R}_s$, we can get the initial eigenvectors $\mathbf{w}_0$ of the selected color pixel. Also, we can transform the initial RGB color pixels $\mathbf{s}_0$ to eigen-space using eigenvectors $\mathbf{w}_0$ as following:

$$\mathbf{y}_0 = \mathbf{w}_0^T \cdot \mathbf{s}_0. \tag{2.32}$$

The term $\mathbf{y}_0$ is projected vectors of the initial color sample, which forms both the signal space and noise spaces. We can iteratively update the next eigenvectors with $\mathbf{y}_0$. The AESS method is illustrated by the following equations as:

$$\mathbf{w}_k' = \mathbf{w}_k \pm 2\mu\mathbf{s}_k\mathbf{y}_k^T. \tag{2.33}$$

where

$$\mathbf{w}_{k+1} = \frac{\mathbf{w}'_k}{||\mathbf{w}_k||}. \tag{2.34}$$

and

$$\mathbf{y}_{k+1} = \mathbf{w}^T_{k+1}\mathbf{s}_{k+1}, \quad k = 0,\dots,N. \tag{2.35}$$

The term $\mathbf{w}'_k$ is the eigenvector with deviation $2\mu\mathbf{s}_k\mathbf{y}^T_k$ belongs to previous pixel and $\mathbf{w}_{k+1}$ is the updated eigenvector of the current pixel. The vector $\mathbf{s}_k$ consists of the gray values of red, green, and blue component of the current pixel. The value of the converging parameter μ is small that approximates to 10^{-6}. The plus sign in (2.33) will conduct the equation to reach a maximum value that represents the signal subspace. The minus sign in (2.33) will find the noise subspace that is converged to a minimum value. For iterative computation of the covariance matrix, $\mathbf{R}_s$, it can be updated as

$$\mathbf{R}_s(k+1) = (1-\alpha)\mathbf{R}_s(k) + \alpha\mathbf{s}_{k+1}\mathbf{s}^T_{k+1}, \tag{2.36}$$

where $\mathbf{R}_s(k)$ represents the kth covariance matrix of the kth sample color pixel and $\mathbf{R}_s(k+1)$ represents the $(k+1)$th covariance matrix of the $(k+1)$th sample color pixel. We can apply following equation to update the eigenvalues without computing the covariance every time.

$$\lambda_{k+1} = (1-\alpha)\lambda_k + \alpha|\mathbf{y}_{k+1}|^2, \tag{2.37}$$

where λ_k represents the kth eigenvalue and λ_{k+1} represents the $(k+1)$th eigenvalue. We can find the mean value of the estimated eigenvectors from (2.33). The estimated mean eigenvectors can be represented as follows:

$$E[\mathbf{w}'_k] = E[\mathbf{I} \pm \mu\mathbf{R}_s(k)]E[\mathbf{w}_k], \tag{2.38}$$

where I is the identity matrix. Inspecting (2.38), we find out that normalized eigenvector $\mathbf{w}'_k$ of previous pixel will approximately equal to eigenvector $\mathbf{w}_k$ of the current pixel. According to this property, we can select μ to adaptively adjust and estimate the eigenvectors of the successive pixels according to (2.33). Using plus and minus sign in (2.38), $\mathbf{w}_k$ will converge to signal plane and noise planes respectively.

2.3.2 Search Algorithms for Desired Color Object

The AESS algorithm is applied to the desired region with first selecting an initial pixel. We take the sampled pixel as the starting point of our searching algorithms. From this starting point, the eigenvectors will be adaptively updated according to different searching routes. Finally, the desired color object will be segmented. The quality of the segmented result is heavily influenced by the shading condition, because the eigenvectors are very sensitive to the color shade. In order to overcome this shading effect, we proposed three searching algorithms to solve the problems. Using this method, we hope that the segmented result will not be influenced by the background with similar color. With adaptively updating eigen-subspaces according to the search routes, the proposed method can obtain a distinct object. The three search algorithms will be described as follows. The square spiral search (SSS) algorithm is shown in Fig. 2.11. The four quadrant search (FQS) algorithm is shown in Fig. 2.12. The slant horizontal vertical search (SHVS) algorithm is shown in Fig. 2.13. The

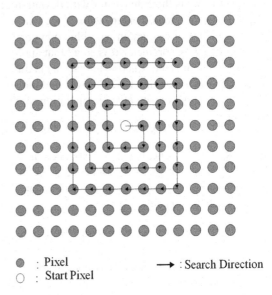

Fig. 2.11 SSS search
algorithm

● : Pixel   → : Search Direction
○ : Start Pixel

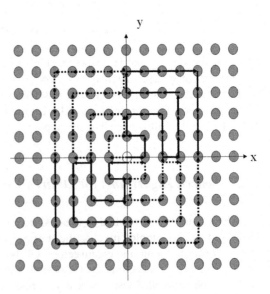

Fig. 2.12 FQS search
algorithm

Fig. 2.13 SHVS search algorithm

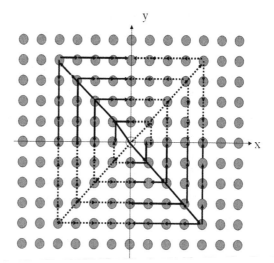

advantage of the SSS method is suitable to segment the interior part of the object because its color is changing smoothly. But, the SSS method is not good at segmenting the exterior part of the object because the boundaries with different color will be a major problem. The FQS method is searching separately in four quadrants where each area is independent. The advantage of the FQS method is that different color area will not influence each other. The disadvantage of the FQS method is that small area in certain quadrant will be difficult to extract. The main reason is that the eigenvectors are updating too fast and is hard to segment the abrupt color-changing area. The SHVS method takes both advantages from the SSS and the FQS methods that is most suitable to our simulation.

2.3.3 Block Diagram of Adaptive Eigen-Subspace Segmentation Method

Figure 2.14 shows the block diagram of the proposed AESS method. First, we will sample the chosen pixel and the surrounding eight pixels to obtain the initial eigenvectors as in Fig. 2.15. Then, new eigenspaces, related to the pixel obtained by the searching algorithms as in Figs. 2.11, 2.12, and 2.13 will be formed. Then, we apply the eigen-subspace transformation of the color planes to form the signal and noise planes as shown in Fig. 2.16. Finally, we can use (2.39) to differentiate between the desired and unwanted color pixel. If the pixel is justified as a desired color pixel then the AESS method will be applied otherwise the algorithm will search next pixel and take the eigenvectors of current pixel as reference. We devise an equation to separate

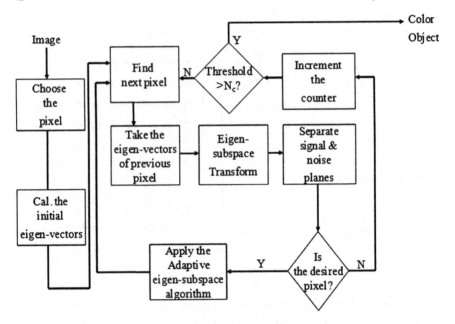

Fig. 2.14 Adaptive eigen-subspace segmentation (AESS) algorithm

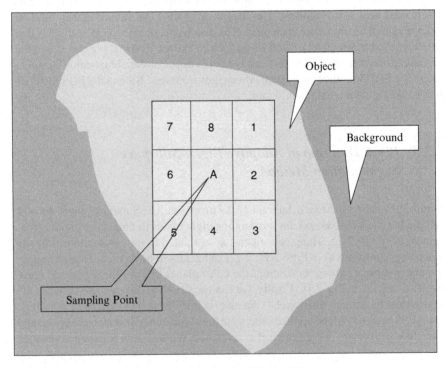

Fig. 2.15 Adaptive eigen-subspace segmentation (AESS) algorithm

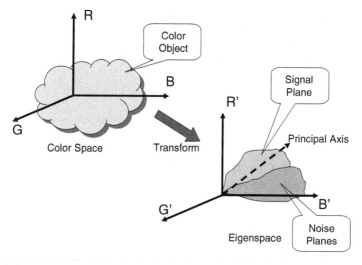

Fig. 2.16 Color space transform to eigenspace

the desired and unwanted pixels that is described as follows:

$$\left| \frac{SA_1^{\alpha}}{NA_1^{\beta} \cdot SA_2^{\gamma}} - \frac{SB_1^{\alpha}}{NB_1^{\beta} \cdot NB_1^{\gamma}} \right| < \Pi, \tag{2.39}$$

where

$$SA_1^{\alpha} = \mathbf{s}_k \cdot \mathbf{w}_{k-1,s}$$

$$NA_1^{\beta} = \mathbf{s}_k \cdot \mathbf{w}_{k-1,n1}$$

$$NA_2^{\gamma} = \mathbf{s}_k \cdot \mathbf{w}_{k-1,n2}$$

$$SB_1^{\alpha} = \mathbf{s}_{k-1} \cdot \mathbf{w}_{k-2,s}$$

$$NB_1^{\beta} = \mathbf{s}_{k-1} \cdot \mathbf{w}_{k-2,n1}$$

$$NB_2^{\gamma} = \mathbf{s}_{k-1} \cdot \mathbf{w}_{k-2,n2}.$$

In (2.39), the first term represents the current pixel and the second term represents the previous pixel. The index k represents the current searching location and $k-1$ represents the previous location. The term $\mathbf{s}_k$ and $\mathbf{s}_{k-1}$ represent the current and previous RGB color pixel. The term $\mathbf{w}_{k-1,s}$, $\mathbf{w}_{k-1,n1}$, and $\mathbf{w}_{k-1,n2}$ represent the eigenvectors of the $(k-1)$th vectors in signal and two noise planes. The terms, $\mathbf{w}_{k-2,s}$, $\mathbf{w}_{k-2,n1}$, and $\mathbf{w}_{k-2,n2}$ represent the eigenvectors of the $(k-2)$th vectors in signal and two noise planes. The term of SA_1 and SB_1 represent the RGB pixels projected on the signal plane, whereas, NA_1, NA_2, NB_1, and NB_2 are the RGB pixels projected on the noise plane. We find out that is appropriate to take $\alpha = 2$, $\beta = 1$, and $\gamma = 1$ in simulation. If the difference value of the projection exceeds the threshold Π, then the searching point is set to gray level $= 255$ otherwise set to 0. Finally, simulation will be terminated if the pixel number of gray $= 0$ has exceeded certain threshold number N_c.

2.3.4 Simulation Results

Figure 2.17a,b show the result of SSS and FQS algorithm, separately. The simulation sequence Mitq, Clair, Bream, and Students are shown in Fig. 2.18a. The desired color objects that are to be segmented are shown as arrow marks. Applying the SHVS algorithm on the four sequences, the segmented results are shown from Fig. 2.18b. Comparing the Mitq segmentation results in Figs. 2.18 and 2.17, it can be shown that the SHVS algorithm is better than the other two search algorithms. Figure 2.19 shows the segmentation results of Bream sequence with two different parameters according to (2.39): (a) $\alpha = 2$, $\beta = 1$, and $\gamma = 1$ and (b) $\alpha = 2$, $\beta = 1$, and $\gamma = 0$.

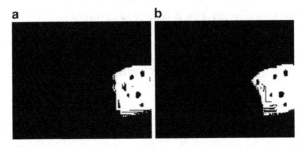

Fig. 2.17 (a) SSS (b) FQS Segmentation Results of Mitq sequence

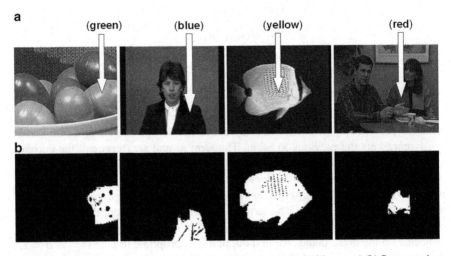

Fig. 2.18 (a) Original sequences (Desired color objects are marked with arrows) (b) Segmentation results by SHVS Methods

Fig. 2.19 Segmentation
results with parameters of
Bream sequence by AESS
method: (**a**) $\alpha = 2$, $\beta = 1$,
$\gamma = 1$; (**b**) $\alpha = 2$, $\beta = 1$, $\gamma = 0$

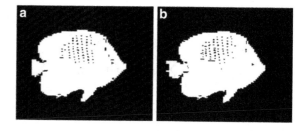

2.4   Color Object Segmentation Using Fuzzy C-Means (FCM) with Eigen-Subspace Projection

In this section, two eigen-based FCM methods by combing both PCT and FCM concepts together to achieve effective color segmentation were proposed. In Sect. 2.4.1, conventional FCM algorithms are briefly reviewed. In Sect. 2.4.2, the separated eigen-based FCM (SEFCM) algorithm with the FCM clustering mechanism is separately applied to projections of signal and noise subspaces. Then, a coupled eigen-based FCM (CEFCM) method by introducing an eigen-based membership function embedded in the FCM cluster process is introduced. In Sect. 2.4.3, the simulation results show to verify the proposed methods for any desired color objects segmentation.

2.4.1   Fuzzy C-Means (FCM)

The fuzzy C-means (FCM) algorithm [30–32] is an iterative unsupervised clustering algorithm that robustly adjusts representative centers of each pattern to best partition the data into several distinct classes. The clustering process is accomplished by minimizing an objective function, which is defined by some measure similarity of the data samples. The objective function can be expressed as follows [30, 31]:

$$J_m(\mathbf{U},\mathbf{V};\mathbf{X}) = \sum_{q=1}^{N}\sum_{j=1}^{c} u_{jq}^m \cdot dist^2(\mathbf{x}_q,\mathbf{v}_j), \qquad (2.40)$$

where N is the number of the data, c is the number of clusters, and scalar m is the arbitrary chosen FCM weighting exponent, which must be greater than one. In (2.40), $\mathbf{X} = \{\mathbf{x}_1,\mathbf{x}_2,\ldots,\mathbf{x}_N\}$ denotes a set of unlabeled column vectors and $\mathbf{V} = \{\mathbf{v}_1,\mathbf{v}_2,\ldots,\mathbf{v}_c\}$ represents the unknown prototypes, which are known as the cluster centers. The vectors $\mathbf{x}_q$ and $\mathbf{v}_j$ are both k-dimensional real Euclidean space $\mathfrak{R}^k$. Hence, the similarity measurement $dist(\mathbf{x}_q,\mathbf{v}_j)$ can be specified as either the Euclidean distance or the Mahalanobis distance. The fuzzy C-partition matrix $\mathbf{U}$ is with size of $c \times N$ that its element can be defined as $u_{j,q} \in M_{\text{fcm}}$ as,

$$M_{\text{fcm}} \equiv \left\{ \mathbf{U} = (u_{\text{jq}}) | 0 \le u_{\text{jq}} \le 1, \textbf{for all } j,q; \right.$$

$$\left. \sum_{j=1}^{c} u_{\text{jq}} = 1, \textbf{for all } q; 0 < \sum_{q=1}^{N} u_{\text{jq}} < N, \textbf{for all } j \right\}.$$

If $dist(\mathbf{x}_q, \mathbf{v}_j)$ is specified as the Euclidean distance then it can be expressed as

$$dist(\mathbf{x}_q, \mathbf{v}_j) = \left[\sum_{\alpha=1}^{k} (x_{q\alpha} - v_{j\alpha}) \right]^{\frac{1}{2}}. \tag{2.41}$$

where $x_{q\alpha}$ and $v_{j\alpha}$ are the elements in the vector of $\mathbf{x}_q$ and $\mathbf{v}_j$. If the distance $dist(\mathbf{x}_q, \mathbf{v}_j)$ is an inner product norm that is called Mahalanobis distance, then, it is expressed as

$$dist^2(\mathbf{x}_q, \mathbf{v}_j) = \|\mathbf{x}_q - \mathbf{v}_j\|^T \mathbf{A}_j \|\mathbf{x}_q - \mathbf{v}_j\| = \mathbf{Q}_j^T \mathbf{A}_j \mathbf{Q}_j. \tag{2.42}$$

In (2.42), $\mathbf{A}_j$ is a $k \times k$ positive defined matrix derived from the jth cluster. When $\mathbf{A}_j = \mathbf{I}$, (2.42) is equal to the Euclidean norm as specified in (2.41). For $m > 1$ and $\mathbf{x}_q \ne \mathbf{v}_j$, the objective function $J_m(\mathbf{U}, \mathbf{V}; \mathbf{X})$ may lead to a minimum if the following equations hold:

$$u_{\text{jq}} = \frac{(dist_{\text{jq}})^{\frac{-2}{m-1}}}{\sum_{i=1}^{c} (dist_{\text{iq}})^{\frac{-2}{m-1}}} \quad \forall \, j, q. \tag{2.43}$$

and

$$\mathbf{v}_i = \frac{\sum_{q=1}^{N} (u_{\text{jq}})^m \mathbf{x}_q}{\sum_{q=1}^{N} (u_{\text{jq}})^m} \quad \forall \, i. \tag{2.44}$$

The similarity measure terms $dist_{\text{jq}}$ and $dist_{\text{iq}}$ specified in (2.43) can be defined as either (2.41) or (2.42) with respective cluster center $\mathbf{v}_i$ or $\mathbf{v}_j$. Unlike traditional classification algorithms, the FCM algorithm assigns all object patterns to each cluster in fuzzy fashions. Each pattern associated with a belonging specified by membership grades between 0 and 1. The fuzzy membership value describes how close or accurate a sample resembles an ideal element of a population. The imprecision caused by vagueness or ambiguity is characterized by the membership value. Inclusive of the concept of fuzziness, the FCM algorithm computes each class center more precisely and with higher robustness to the noise. The procedures of the FCM algorithm [30, 31] are enlisted as follows:

1. Initialization: Fix the number of cluster c and feature coefficient m, set iteration loop index $t = 0$, and select initial cluster centers.
   We are randomly select c initial cluster centers from the space as $\mathbf{v}_j^{(0)}$, for $j = 1, 2, \ldots, c$. Initialized $\mathbf{U}^{(0)}$.
2. Sampling: Choose total N data samples $\mathbf{x}_q$ for $q = 1, 2, \ldots, N$ from the image. It is performed by clicking the mouse on the image.

3. Calculating the fuzzy cluster centers: Compute all cluster centers $\{\mathbf{v}^{(t)}\}$ using $\mathbf{U}^{(t-1)}$ with the equation specified in (2.9).
4. Update membership function $\mathbf{U}^{(t)}$: Update $\mathbf{U}^{(t)}$ using $\mathbf{v}^{(t)}$ with the equation specified in (2.8).
5. Check convergence condition: Check the previous defined convergence behavior as $\triangle$ by computing

$$\triangle = \left| \mathbf{v}^{(t)} - \mathbf{v}^{(t-1)} \right|. \tag{2.45}$$

6. If $\triangle < \varepsilon$ or a preset loop count N_t is reached then terminate; otherwise set $t = t+1$ and go to Step 3, where ε is the preset terminating criterion.

In (2.45), the superscript t denotes the number of iterations. If the changes of the class centers are less than a predefined criterion ε that means the objective function $J_m(\mathbf{U}, \mathbf{V}; \mathbf{X})$ is no longer decreasing. The final segmentation result is achieved.

To improve orientation sensitivity (OS), Schmid in [33] suggested a modified FCM algorithm (OSFCM) by modifying $\mathbf{A}_j$ described in (2.42) as:

$$\mathbf{A}_j = \mathbf{V}_j^T \mathbf{L}_j \mathbf{V}_j, \tag{2.46}$$

where $\mathbf{L}_j$ denotes the diagonal matrix containing the inverse of eigenvalues and $\mathbf{V}_j$ represents the unitary matrix lining up the corresponding eigenvectors of the fuzzy covariance matrix $\mathbf{C}_j^x$ for the jth cluster. The fuzzy covariance matrix for the jth cluster $\mathbf{C}_j^x$ is given by

$$\mathbf{C}_j^x = \frac{1}{N} \sum_{q=1}^{N} u_{jq}^m \left(\mathbf{x}_q \mathbf{x}_q^T - \mathbf{v}_j \mathbf{v}_j^T \right). \tag{2.47}$$

From simple matrix derivations, it is obvious that $\mathbf{A}_j = (\mathbf{C}_j^x)^{-1}$.

2.4.2   Eigen-Based FCM Algorithms

In this section, we combine the FCM classification with eigen-subspaces projection together to achieve effective color segmentation. By using eigenvectors, we can transform the original color space into the modal coordinate system of the desired color as

$$\mathbf{z}_q = [\mathbf{w}_q \ \mathbf{w}_2 \ \mathbf{w}_3]^T \mathbf{x}_q = [\phi_q \ \varphi_q \ \psi_q]. \tag{2.48}$$

Now, the first-principal elements, ϕ_q for $q = 1, 2, \ldots, N$ specify the signal subspace whereas the second and the third elements φ_q and ψ_q and build the noise subspaces. The vector $\mathbf{x}_q$ is defined as in (2.40). With signal and noise subspaces, we develop two eigen-based FCM detection procedures, the separate eigen-based FCM (SEFCM) and the coupled eigen-based FCM (CEFCM) methods. The main procedures of eigen-based FCM are shown in Fig. 2.20. First, we compute the covariance matrix $\mathbf{R}_s$ of the desired color samples from the RGB color planes followed by

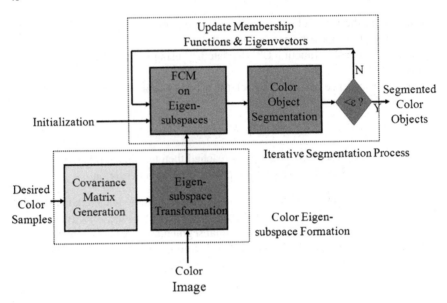

Fig. 2.20 The proposed eigen-based FCM algorithms

the eigen-analysis procedures to obtain the eigenvectors. Then, we perform the eigen-subspace transformation of image color planes with (2.48). Finally, the iterative segmentation process with updated membership functions will be applied on the eigen-subspaces. Although the eigenvectors are generated by the selected color samples, we still need to adjust the eigen-subspaces to achieve more satisfactory segmentation results. We are iterative in adjusting the eigen-subspaces with new eigenvectors obtained from new covariance matrix. The detail algorithms of SEFCM and CEFCM will be addressed in Sects. 2.4.2.1 and 2.4.2.2.

2.4.2.1 Separate Eigen-Based FCM (SEFCM) Method

In this section, we have proposed the SEFCM algorithm to separately consider the signal and noise planes. During simulation, we iteratively construct new covariance matrices that are similar to (2.47) by using the eigen-subspace data z_q instead of x_q in color image. The expression of the covariance matrix can be stated as follows:

$$\mathbf{C}_j^z = \frac{1}{N} \sum_{q=1}^{N} u_{jq}^m \mathbf{z}_q \mathbf{z}_q^T. \qquad (2.49)$$

The color objects will be extracted after the objective function reaches a minimum. With the help of the eigenvalues, we can obtain the represented segmented color objects with respective to the signal and noise subspaces. Following, with a simple logical "AND" operation on both results, we can obtain the segmentation of the desired video objects correctly. In the SEFCM, we modify the matrix $\mathbf{L}_j$ as in (2.46) that is suitable to extract the signal and noise subspaces. For extracting the signal space, we can rewrite $\mathbf{L}_j$ as follows:

$$
\Gamma_j = \begin{pmatrix} \left(\frac{\lambda_{2,j}+\lambda_{3,j}}{2}\right)^{-1} & 0 & 0 \\ 0 & \lambda_{1,j}^{-1} & 0 \\ 0 & 0 & \lambda_{1,j}^{-1} \end{pmatrix}. \tag{2.50}
$$

Similarly, we can extract the noise planes by using the following matrix:

$$
\Gamma_j = \begin{pmatrix} \lambda_{1,j}^{-1} & 0 & 0 \\ 0 & \left(\frac{\lambda_{2,j}+\lambda_{3,j}}{2}\right)^{-1} & 0 \\ 0 & 0 & \left(\frac{\lambda_{2,j}+\lambda_{3,j}}{2}\right)^{-1} \end{pmatrix}. \tag{2.51}
$$

We adopt (2.50) to extract the signal plane by using $\lambda_{1,j}^{-1}$ to suppress the noise terms. In (2.51), we use $\lambda_{1,j}^{-1}$ to suppress the signal terms in order to obtain two noise planes. We can modify the membership function of (2.43) as follows:

$$
u_{jq} = \frac{\left[(\mathbf{z}_q - \mathbf{v}_j)^T \mathbf{A}_j (\mathbf{z}_q - \mathbf{v}_j)\right]^{\frac{-2}{m-1}}}{\sum_{\beta=1}^{c} \left[(\mathbf{z}_q - \mathbf{v}_\beta)^T \mathbf{A}_\beta (\mathbf{z}_q - \mathbf{v}_\beta)\right]^{\frac{-2}{m-1}}}, \tag{2.52}
$$

where $\mathbf{A}_j = \mathbf{V}_j^T \Gamma_j \mathbf{V}_j$ and $\mathbf{A}_\beta = \mathbf{V}_\beta^T \Gamma_\beta \mathbf{V}_\beta$ related to class j and β, respectively. For class β, the index j appeared in (2.50) and (2.51) should changed to β. The detailed procedures of the SEFCM are shown in Fig. 2.19 and illustrates as follows:

1. Sample few desired color object blocks.
2. Compute the covariance matrix and obtain the eigenvectors according to (2.2).
3. Transform the color images to signal and noise subspaces with eigenvectors as (2.48).
4. Initialize the modified membership value and center of each cluster. With iterative updating of the covariance matrices using (2.49), apply FCM to extract the segmentation results related to signal and noise planes separately. Either segmenting on signal or noise planes, we apply (2.50) or (2.51) to the new membership function (2.52) during the FCM classification procedures.
5. Perform logical operation on the results obtained from Step 4.

2.4.2.2   Coupled Eigen-Based FCM (CEFCM) Method

In order to efficiently segment the desired color objects, we devise a coupled eigen-based FCM (CEFCM) algorithm (Fig. 2.21). In considering signal and noise planes

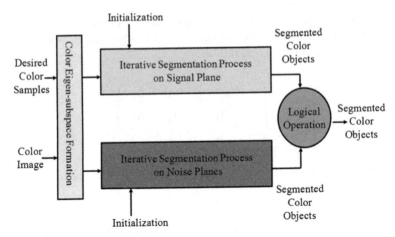

Fig. 2.21 Signal flow diagram of proposed SEFCM algorithm

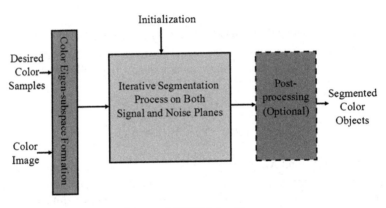

Fig. 2.22 Signal flow diagram of proposed CEFCM algorithm

together, the CEFCM adopts three dimensional eigen-subspaces data for classification. The function block diagram of the CEFCM is shown in Fig. 2.22. Similar to SEFCM, we also construct new covariance matrices that are similar to (2.47) by using the eigen-subspace data $\mathbf{z}_q$ in stead of $\mathbf{x}_q$ in color images. In view of statistical inference and fuzzy property, we can construct a new covariance matrix for the jth cluster center as

$$\hat{\mathbf{C}}_j^z = \frac{1}{\sum_{q=1}^{N} u_{jq}^m} \sum_{q=1}^{N} u_{jq}^m \mathbf{z}_q \mathbf{z}_q^T. \tag{2.53}$$

It is not necessary to iteratively rebuild the covariance matrix and construct the new eigen-subspaces from the original color images because the color objects are selected under our inspection. We can gradually adjust the direction of principal axes by using the already built eigen-subspaces so that large amount of transformation computations can be saved. In updating procedures, we adopt the new covariance matrix as described in (2.53), where the belongings are treated as the weightings of $\mathbf{z}_q\mathbf{z}_q^T$ that is formed by the data in the eigen-subspaces. From covariance matrix $\hat{\mathbf{C}}_j^z$, we can obtain class j's eigenvector $\mathbf{w}_{j,i}$ and the corresponding eigenvalue $\lambda_{j,i}$, where index i denotes the ith component because we order the eigenvalues as

$$\lambda_{j,1} \geq \lambda_{j,2} \geq \lambda_{j,3}. \tag{2.54}$$

Then, the jth cluster center of signal term in the CEFCM can be expressed by the principal component

$$\mathbf{v}_{j1} = \sqrt{\lambda_{j,1}}\mathbf{w}_{j,1}. \tag{2.55}$$

The second and the third components are treated as the noise terms. The similarity measure related to the jth cluster center expressed as Euclidean distance between $\mathbf{z}_q$ and $\mathbf{v}_{j,1}$ now becomes the orthogonal projection to the noise eigenvectors. The smaller of $\|\mathbf{z}_q - \mathbf{v}_{j,1}\|$ means that $\mathbf{z}_q$ and $\mathbf{v}_{j,1}$ are closer to each other. Based on eigen properties, the smaller projection to both $\mathbf{w}_{j,2}$ and $\mathbf{w}_{j,3}$ with respect to $\sqrt{\lambda_{j,2}}$ and $\sqrt{\lambda_{j,3}}$ indicates that $\mathbf{z}_q$ and $\mathbf{v}_{j,1}$ are closer. The measure of $\|\mathbf{z}_q - \mathbf{v}_{j,1}\|$ related to $\mathbf{w}_{j,2}$ and $\mathbf{w}_{j,3}$ is equivalent to the projection of $\mathbf{z}_q$ to the normalized noise eigenvectors, which can be expressed as $\frac{1}{\sqrt{\lambda_{j,2}}}\mathbf{w}_{j,2}$ and $\frac{1}{\sqrt{\lambda_{j,3}}}\mathbf{w}_{j,3}$, so that the membership of the qth sample can be modified as follows:

$$u_{jq} = \frac{\left(\frac{1}{\lambda_{j,2}}\|\mathbf{z}_q^T\mathbf{w}_{j,2}\|^2 + \frac{1}{\lambda_{j,3}}\|\mathbf{z}_q^T\mathbf{w}_{j,3}\|^2 + \frac{1}{\lambda_{j,1}}\left(\|\mathbf{z}_q^T\mathbf{w}_{j,1}\|^2 - \lambda_{j,1}\right)\right)^{\frac{-2}{m-1}}}{\sum_{\beta=1}^{c}\left(\frac{1}{\lambda_{\beta,2}}\|\mathbf{z}_q^T\mathbf{w}_{\beta,2}\|^2 + \frac{1}{\lambda_{\beta,3}}\|\mathbf{z}_q^T\mathbf{w}_{\beta,3}\|^2 + \frac{1}{\lambda_{\beta,1}}\left(\|\mathbf{z}_q^T\mathbf{w}_{\beta,1}\|^2 - \lambda_{\beta,1}\right)\right)^{\frac{-2}{m-1}}}. \tag{2.56}$$

The success of extracting video objects depends on the proportion of three eigenvalues. Inspecting (2.56), we have adjusted the iterating processes near the cluster center in the signal subspace according to its eigenvalue to eliminate too bright or too dark circumstances. In our experiments, we take fuzzy weighting $m = 3$, the class number equals to 6 and feature number equals to 3. Observing the simulation results, we can obtain more satisfactory ones by just considering the principal plane and the strongest noise plane. In this case, we set $\lambda_{j,1} = \lambda_{j,3} = 1$, $\lambda_{\beta,1} = \lambda_{\beta,3} = 1$, and $\lambda_{j,2} \to \infty$; $\lambda_{\beta,2} \to \infty$ with $c = 2$ in (2.56).

After obtaining the segmentation results, almost all the desired color pixels can be found. The few remaining noise pixels can be easily removed by any post-processing procedure. The detailed procedures for the CEFCM are listed as follows:

1. As Step 1 in Sect. 2.4.2.1.
2. As Step 2 in Sect. 2.4.2.1.
3. As Step 3 in Sect. 2.4.2.1.
4. Initialize the membership function and cluster centers. Later, we jointly consider three eigen-subspaces and iteratively update the covariance matrix with (2.53).

With newly found eigenvectors and eigenvalues, we apply (2.55) and (2.56) to perform the FCM classification processes.

5. (Optional). Post-processing procedure is optional for smoothing the results from Step 4. In our experiments, we do not intent to use any post-processing procedures in order to show the inherent classified capability of this algorithm.

2.4.3   Simulation Results

We directly apply the conventional FCM to these four sequences in the RGB color planes. The simulation results are shown in Fig. 2.23. Without any threshold determination, the desired objects obtained by the traditional FCM are better than those obtained by the PCT method. However, the segmented results still contain many unwanted noises.

Without any other assistance, Fig. 2.24 shows the simulation results by applying conventional FCM to the transformed planes, which are performed by the KL projections. Compared to Fig. 2.23, some improvements are achieved. The KL projections obtained from the desired color samples can translate the image data to the desired working space in more compaction form. It is reasonable to apply the segmentation efforts on the eigen-subspaces. Theoretically, results of Figs. 2.6 and 2.24 should be identical because of the linear transformation between color-space and eigenspace. The difference of the results shown in these two figures may be due to initial data distribution and class centers.

For comparison, we also apply OSFCM algorithm [33] to the eigen-subspaces. Figure 2.25 shows the segmented images obtained from the OSFCM method. Although all the main objects can be detected, the objects with near color are also

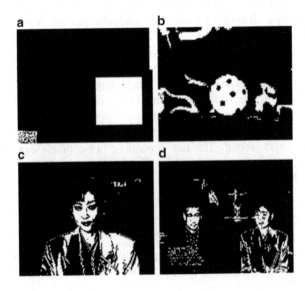

Fig. 2.23 Segmented images obtained by the conventional FCM directly applying on R,G,B planes: (a) Mosaic; (b) Ball; (c) Akiyo; (d) News sequences

Fig. 2.24 Segmented images obtained by the conventional FCM applying on the KL transformed color spaces: (**a**) Mosaic; (**b**) Ball; (**c**) Akiyo; (**d**) News sequences

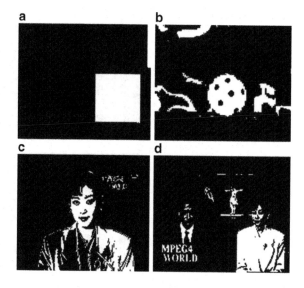

Fig. 2.25 Segmented images obtained by the OSFCM method: (**a**) Mosaic; (**b**) Ball (**c**) Akiyo; (**d**) News sequences

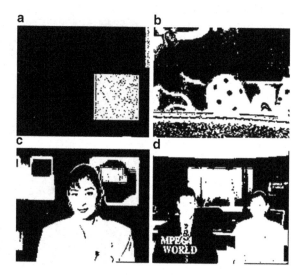

extracted. Similar to the PCT approach, the OSFCM method does not separately use the noise-subspace. The color objects with high signal-subspace projects will be erroneously included in the similar color pixels.

Figure 2.26 shows the results obtained by the SEFCM. We can find out that most of the noise has been removed compared to the previous obtained results. The major defeat of the SEFCM method appears on the clothes of Akiyo since both signal and noise subspace projections do not perform simultaneously. In order to improve the performance, the CEFCM algorithm adopts the signal and noise subspace

Fig. 2.26 Segmented images obtained by the SEFCM method: (**a**) Mosaic; (**b**) Ball (**c**) Akiyo; (**d**) News sequences

Fig. 2.27 Segmented images obtained by the CEFCM method: (**a**) Mosaic; (**b**) Ball; (**c**) Akiyo; (**d**) News sequences

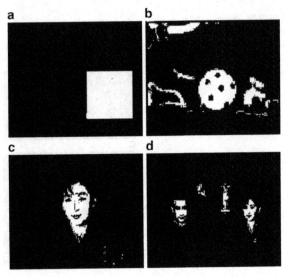

projections together. Figure 2.27 shows the segmented images obtained from the CEFCM method. It is obvious that the CEFCM method outperforms the other color object segmentation algorithms.

2.5 Conclusions

Video object segmentation has been recognized as a main technology to achieve the content-based coding proposed in the MPEG-4 standard. In this chapter, we proposed color eigenspace segmentation methods to extract the desired objects.

After theoretical analyses, the color eigen-structure segmentation algorithm uses both signal and noise subspaces effectively. Without any pre-processing process, we can precisely detect the desired color and preserve small significant features. If the segmented results still contain some undesirable pixels, of course, we can easily remove them once we further introduce temporal or motion information. The proposed color segmentation algorithm performs successfully for single color objects. When the desired object is with more than two colors, for example color texture or color pattern, we should apply our algorithm several times accordingly. The final video object planes will be the union of all the segmented results.

In Sect. 2.3, adaptive eigen-subspace segmentation (AESS) algorithm to locally extract the desired objects has proposed. Three search methods are used in the AESS algorithm which is SSS, FQS, and SHVS. Using this algorithm, we can extract the desired color objects even if the background has the same color as the object. The simulation results show that the AESS algorithm can achieve better segmentation results. It has a localized segmentation capability.

In Sect. 2.4, we use the conventional FCM combined with the eigenspace projections concept to develop two segmentation algorithms. The first algorithm uses the eigen-structure combined with the FCM method and the second algorithm simply applies threshold on the eigen-spaces according to the statistical properties. The method with FCM is very effective to segment desired objects without considering any threshold. The drawback of the first method is that it costs large amount of computation time. The second method adopting statistical analysis is faster than the first one but needs to select different threshold values according to different sequences.

Considering signal and noise projections, the SEFCM method shows its effectiveness comparing to use the PCT or the FCM algorithm alone. In order to achieve satisfactory simulation results, we further suggest the CEFCM algorithm to improve the segmentation performance. Compared to the conventional FCM method and the OSFCM method, we found that the SEFCM and CEFCM achieve the best segmentation performance, which is robust and less susceptible to the noise. Further integrated with spatial and temporal information, our proposed algorithms can achieve even better results in the future works. The difference between AESS algorithm and the other method is that our proposed method can be used to segment the desired object at will. Especially, when several identical color objects exist at the same frame, our algorithm can extract only the designated object out. Using color eigen-subspace to segment the object has been proven to be an effective method but it maybe failed when the color shade is present. In addition to the normal color distribution conditions, our algorithm is most suitable to segment the object with color shade.

References

1. ISO/IEC JTC1/SC29/WG11, N2201 (1998) Text for ISO/IEC FCD 14496-1 Systems
2. ISO/IEC JTC1/SC29/WG11, N2502 (1998) Information technology - coding of audio-visual objects, Part 2: visual

3. Lee MC, Chen W, Lin CB, Gu C, Markoc T, Zabinsky SI, Szeliski R (1997) A Layer Video Object Coding System Using Sprite and Affine Motion Model, IEEE Transactions on Circuits and Systems for Video Technology, vol. 7, no. 1, pp. 130–145.
4. Zheng H, Blostein SD (1995) Motion-based object segmentation and estimation using the MDL principle, IEEE Transactions on Image Processing, vol. 4, no. 9, pp. 1223–1235.
5. Chang MM, Tekalp AM, Sezan MI (1997) Simultaneous motion estimation and Segmentation, IEEE Transactions on Image Processing, vol. 6, no. 9, pp. 1326–1333.
6. Chu CC, Aggarwal JK (1993) The integration of image segmentation maps using region and edge information, IEEE Transactions on Pattern Analysis and Machine Intelligence, vol. 15, no. 12, pp. 1241–1252.
7. Wani MA, Batchelor BG (1994) Edge-region-based segmentation of range images, IEEE Transactions on Pattern Analysis and Machine Intelligence, vol. 16, no. 3, pp. 314–319.
8. Kaup A, Aach T (1998) Coding of segmented images using shape-independent basis functions, IEEE Transactions on Image Processing, vol. 7, no. 7, pp. 937–947.
9. Williams PS, Alder MD (1996) Generic texture analysis applied to newspaper segmentation, IEEE International Conference on Neural Networks, vol. 3, no.3–6, pp. 1664–1669.
10. Waked B, Bergler S, Suen CY, Khoury S (1998) Skew detection, page segmentation, and script classification of printed document images, IEEE International Conference on Systems, Man, and Cybernetics, vol. 5, no. 11–14, pp. 4470–4475.
11. Littmann E, Ritter H (1997) Adaptive color segmentation-a comparison of neural and statistical methods, IEEE Transactions on Neural Networks, vol. 8, no. 1, pp. 175–185.
12. Liu J, Yang YH (1994) Multiresolution color image segmentation, IEEE Transactions on Pattern Analysis and Machine Intelligence, vol. 16, no. 7, pp. 689–700.
13. Shafarenko L, Petrou H, Kittler J (1998) Histogram-based segmentation in a perceptually uniform color space, IEEE Transactions on Image Processing, Vol.7, No. 9, pp. 1354–1358.
14. Gonzalez RC, Woods RE (1992), Digital Image Processing, MA: Addison-Wesley
15. Murase H, Nayar SK (1994) Illumination panning for object recognition using parametric eigenspaces, IEEE Transactions on Pattern Analysis and Machine Intelligence, vol. 16, no. 12, pp. 1219–1227.
16. Xiuping J, Richards JA (1999) Segmented principal components transformation for efficient hyperspectral remote-sensing image display and classification, IEEE Transactions on Geoscience and Remote Sensing, vol. 37, no.1, Part-2, pp. 538–542.
17. Kerfoot IB, Bresler Y (1997) Theoretical analysis of multispectral image segmentation criteria, IEEE Transactions on Image Processing, vol. 8, no. 6 , pp. 798–820.
18. Kumaresan R, Tufts DW (1983) Estimating the angles of arrival of multiple plane waves, IEEE Transactions on Aerospace and Electronic Systems,vol. AES-19, no. 1, pp. 134–139.
19. Hu B, Gosine RG (1997) A new eigenstructure method for sinusoidal signal retrieval in white noise: estimation and pattern recognition, IEEE Transactions on Signal Processing, vol. 45, no. 12, pp. 3073–3083.
20. Kaveh M, Barabell AJ (1986) The statistical performance of the MUSIC and the minimum-norm algorithms resolving plane waves in noise, IEEE Transactions on Acoustics, Speech, and Signal Processing, vol. ASSP-34, no. 2, pp. 331–341.
21. Xu XX, Hao SS, Yang JF (2000) Video object color segmentation using adaptive eigensubspace, Proceedings of 2000 workshop on internet distributed systems, May 11–12, National Cheng Kung University, Tainan, Taiwan, pp. 122–127.
22. Yang JF, Hao SS, Chung PC (2002) Color image segmentation using fuzzy C-means with eigen-subspace projections, Vol. 82, Signal Processing, pp. 461–472.
23. Wan X, Kuo CCJ (1998) A New Approach to Image Retrieval with Hierarchical Color Clustering, IEEE Transactions on Circuits and Systems for Video Technology, vol. 8, no. 5, pp. 628–643.
24. Antoszczyszyn PM, Hannah JM, Grant PM (1998) Reliable tracking of facial features in semantic-based video coding, IEE Proceedings-Vision, Image and Signal Processing, vol. 145, no. 4, pp. 257–263.
25. Special Issue on Very Low Bit Rate Video Coding (1994) IEEE Transactions on Circuits and Systems for Video Technology, vol. 4, no. 3.

26. Cai J, Goshtasby A, Yu C (1998) Detecting human faces in color images, Proceedings of the International Workshop on Multi-Media Database Management Systems, pp. 124–131, OH, USA.
27. Chai D, Ngan KN (1998) Locating facial region of a head-and-shoulders color image, Proceedings of the Third IEEE International Conference on Automatic Face and Gesture Recognition, pp. 124–129, WA, Australia,.
28. Hartung J, Jacquin A, Pawlyk JS, Rosenberg J, Okada H, Crouch PE (1998) Object-Oriented H.263 Compatible Video Coding Platform for Conferencing Applications, IEEE Journal on Selected Areas in Communications, vol. 16, no. 1, pp. 42–55.
29. Lee JH, Chang BH, Kim DS (1994) Comparison of colour transformations for image segmentation, Electronics Letters, vol. 30, no. 20, pp. 1660–1661.
30. Bezdek JC (1981) Pattern recognition with fuzzy objective function algorithm, Plenum, New York.
31. Tsoukalas LH, Uhrig RE (1997) Fuzzy and Neural Approaches in Engineering, John Wiley & Sons, Inc.
32. Haykins S (1999) Neural Networks-A Comprehensive Foundation, 2nd edition, Prentice Hall International, Inc.
33. Schmid P (1999) Segmentation of digitized dermatoscopic images by two-dimensional color clustering, IEEE Transactions on Medical Imaging, Vol. 18, No. 2, pp. 164–171.

Chapter 3
Semantic Object Segmentation

Xiaogang Wang

Abstract Semantic object segmentation is to label each pixel in an image or a video sequence to one of the object classes with semantic meanings. It has drawn a lot of research interest because of its wide applications to image and video search, editing and compression. It is a very challenging problem because a large number of object classes need to be distinguished and there is a large visual variability within each object class. In order to successfully segment objects, local appearance of objects, local consistency between labels of neighboring pixels, and long-range contextual information in an image need to be integrated under a unified framework. Such integration can be achieved using conditional random fields. Conditional random fields are discriminative models. Although they can learn the models of object classes more accurately and efficiently, they require training examples labeled at pixel-level and the labeling cost is expensive. The models of object classes can be learned with different levels of supervision. In some applications, such as web-based image and video search, a large number of object classes need to be modeled and therefore unsupervised learning or semi-supervised learning is preferred. Therefore some generative models, such as topic models, are used in object segmentation because of their capability to learn the object classes without supervision or with weak supervision of less labeling work. We will overview different technologies used in each step of the semantic object segmentation pipeline and discuss major challenges for each step. We will focus on conditional random fields and topic models, which are two types of frameworks widely used in semantic object segmentation. In video segmentation, we summarize and compare the frameworks of Markov random fields and conditional random fields, which are the representative models of the generative and discriminative approaches respectively.

X. Wang (✉)
Department of Electronic Engineering, The Chinese University of Hong Kong,
Hong Kong, China
e-mail: xgwang@ee.cuhk.edu.hk

K.N. Ngan and H. Li (eds.), *Video Segmentation and Its Applications*,
DOI 10.1007/978-1-4419-9482-0_3, © Springer Science+Business Media, LLC 2011

3.1 Introduction

The task of semantic object segmentation is to label each pixel in an image or a video sequence to one of the object classes with semantic meanings (see examples in Fig. 3.1). The object classes can be predefined or unsupervised learned from a collection of images or videos. It is different than unsupervised image and video segmentation, which is to group pixels into regions with homogeneous color or texture but without semantic meanings. It has important applications to image and video search, editing, and compression. For example, semantic regions with their 2D spatial arrangement sketched by users can be used as query to retrieve image. Segmented objects can be deleted from images or copied between images. Different regions of images can be enhanced in different ways based on their semantic meanings.

Semantic object segmentation is a very challenging problem, because there are a very large number of object classes to be distinguished, some object classes are visually similar, and each object class may have very large visual variability. These object classes can be structured, such as cars and airplanes, or unstructured, such as grass fields and water. Due to variations of viewpoints, poses, illuminations, and occlusions, objects of the same class have different appearance across images. In order to develop a successful semantic object segmentation algorithm, there are three important factors to be considered: local appearance, label consistency between

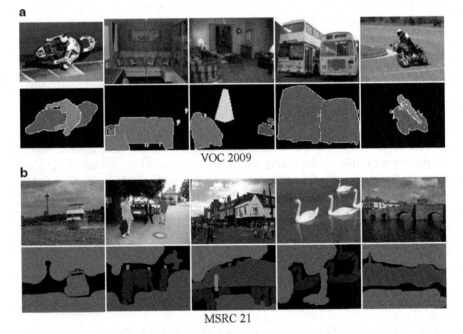

Fig. 3.1 Examples of images (*first row*) and manually segmented objects (*second row*) from PASCAL VOC 2009 [1] (**a**) and MSRC 21 [2] (**b**). Different colors represent object categories

neighboring pixels, and long-range contextual information [3]. In order to model the local appearance at a pixel, filter-banks and visual descriptors are applied to the neighborhood around the pixel and their responses are used as the input of a classifier to predict the object label. The filter-banks, visual descriptors and classifiers have to be carefully designed in order to achieve a good balance between high discriminative power and invariance to noise, clutters, and the changes of viewpoints and illuminations. In order to obtain smooth segmentation results, the label consistency between neighboring pixels needs to be considered. In order for the segmentation to be consistent with the boundaries of objects, the algorithm should encourage two neighboring pixels to have the same object label if there is no strong edge between them. In addition to smoothness, the likelihood of two object classes being neighbors should also be considered for local consistency. For example, it is more likely for a cup to be on the top of desk than on a tree. Only considering the appearance of an image patch leads to ambiguities when deciding its class label. For example, a flat white patch could be from a wall, a car or an airplane. The long-range contextual information of the image may help to solve the ambiguities to some extent. For example, some object classes such as horses and grass are more likely to coexist in the same images. If it is known that the image is an outdoor scene, it is more likely to observe sky, grass, and cars than computers, desks, and floors in that image. Local appearance, local consistency, and long-range contextual can be incorporated in a Conditional Random Field (CRF) model [4], which has been popularly used in semantic object segmentation.

The approaches of semantic object segmentation can be supervised or unsupervised. The supervision at the training stage can be at three different levels:

- Pixel-level: each pixel in an image is manually labeled as one of the object classes.
- Mask-level: an object in an image is located by a bounding box and assigned to a object class.
- Image-level: annotate object classes existing in an image without locating or segmenting objects.

Most discriminative object segmentation approaches including CRF need pixel-level or mask-level labeling for training. They can learn the models of object classes more accurately and efficiently. However, as the fast increase of images and videos in many applications such as web-based image and video search, there are an increasing number of object classes to be modeled. The workload of pixel-level and mask-level labeling is heavy and impractical for a very large number of object classes. In recent years, some generative models, such as topic models borrowed from language processing, have become popular in semantic object segmentation. They are able to learn the models of object classes from a collection of images and videos without supervision or supervised by data labeled at the image-level, whose labeling cost is much less. It is also possible for CRF and topic models to integrate the strengths of both types of approaches.

A typical pipeline of semantic object segmentation is shown in Fig. 3.2. Filter-banks or visual descriptors are first applied to images to capture the local appearance

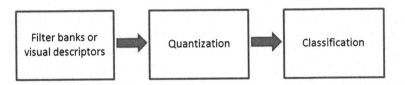

Fig. 3.2 Typical steps of semantic object segmentation. They are done over image pixels, patches or oversegmented superpixels

objects. Their responses are typically quantized into textons or visual words according to codebooks learned in a supervised or unsupervised way. The histograms of textons or visual words are used as input to a classifier to predict labels of object classes. In order to well capture the local consistency and long-range contextual information, CRF or generative models are used to incorporate with local classifiers. These steps can be on at image pixels, patches, or oversegmented superpixels. Many different technologies have been developed to improve each of the three steps. We will review these technologies and discuss the major challenges for these steps. In recent years, some benchmark databases, such as PASCAL VOC 2007 [5], PASCAL VOC 2008 [6], PASCAL VOC 2009 [1], LabelMe [7], LHI [8], and MSRC 21 [2], were published to evaluate the performance of different semantic object segmentation approaches.

In video segmentation, Markov random fields (MRFs) and CRFs are two main frameworks. Statistically, video segmentation formulizes and maximizes a posterior probability of the labels given by the observation data. In the case that there is no or only small number of labeled data, some heuristic or prior knowledge based distributions can be selected to describe the observation data. Based on the selected distributions and the prior of labels modeled in a MRF, the MRF approaches formulate the posterior via likelihoods and priors in Baye's rule. On the contrast, CRFs model the posterior directly to improve the predictive performance if there are large quantities of training data. In CRFs, the model of the observation data is obtained by learning from the training data using some classifiers. Compared to MRFs, CRFs relax the assumption of data independence, while large more expensive labeled data is necessary in CRFs.

This chapter is organized as follows. Section 3.2 introduces different types of filter-banks and visual descriptors to capture local appearance, and different techniques to quantize their responses into textons or visual words. Some popular classifiers on local appearance are reviewed in Sect. 3.3.1. Section 3.3.2 introduces CRF and different approaches of using CRF for semantic object segmentation. Section 3.4 first introduces two classical topic models, Probabilistic Latent Semantic Analysis [9] (pLSA) and Latent Dirichlet Allocation [10] (LDA), which were directly borrowed from language processing and applied to semantic object segmentation. Both pLSA and LDA ignored the spatial distribution of image patches. Spatial Latent Dirichlet Allocation [11], which is an extension of LDA and other topic models incorporating spatial structures of objects are introduced in Sects. 3.4.2 and 3.4.3. The approaches of object segmentations in videos are discussed in Sect. 3.5. Finally the summary is given in Sect. 3.6.

3.2   Local Visual Cues

3.2.1   Filter-Banks and Visual Descriptors

Filter-banks and visual descriptors are used to capture the local appearance of objects. They are calculated from the neighbor of a pixel. On the one hand, they need to be discriminative enough to distinguish a large number of object classes, some of which are visually similar; on the other hand, they need to have invariance to noise, clutters and changes of illuminations and viewpoints. If they are computed at every pixel, computational efficiency is another issue to be considered. In this section we will review some popularly used filter-banks and visual descriptors.

Filter-banks. Filter-banks capture certain frequencies within a neighborhood. Winn et al. [2] proposed a set of filter-banks after testing different combinations of Gaussians, Laplacian of Gaussians (LoG), first and second order derivatives of Gaussians and Gabor kernels on semantic object segmentation. The proposed set of filter-banks included three Gaussians, four LoGs, and four first-order derivatives of Gaussians. The three Gaussian kernels with different standard deviation parameters $\sigma = 1, 2, 4$ were applied to each CIE L,a,b channel. The four LoGs(with $\sigma = 1, 2, 4, 8$) and the four first order derivatives of Gaussians (with $\sigma = 1, 2, 4, 8$) were applied to L channel only. The first order derivatives of Gaussians were in x and y directions. See the kernels of the proposed filter-banks in Fig. 3.3. Some other filter-banks, such as rotation-invariant filters and maximum-response filters, were also proposed [12–14]. A comparison study can be found in [15].

SIFT. SIFT (Scale-Invariant Feature Transform) (see Fig. 3.4) proposed by Lowe [16] is the most widely used local visual descriptors. It has reasonable invariance to changes in illumination, rotation, scaling, and small changes in viewpoints. SIFT keypoints were detected by finding local extrema of Difference-of-Gaussian (DoG)

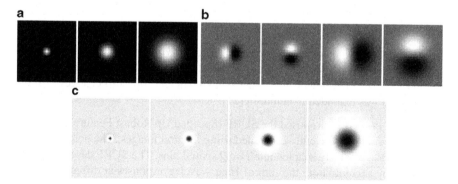

Fig. 3.3 A set of filter banks proposed by Winn [2]. (**a**) Three Gaussian kernels with $\sigma = 1, 2, 4$. They were applied to each CIE L,a,b channel. (**b**) Four derivatives of Gaussians divided into the x- and y-aligned sets, each with two different values of $\sigma = 2, 4$. They were applied to L channel. (**c**) Four Laplacian of Gaussians with $\sigma = 1, 2, 4, 8$. They were applied to L channel

Fig. 3.4 SIFT descriptor [16]
is computed by combining
the normalized orientation
histograms of gradients
within subregions of the
keypoint into a feature vector

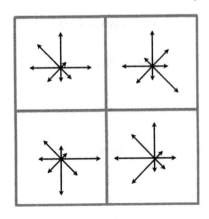

filters at different scales. For each keypoint, its orientation and scale were selected. A SIFT descriptor of a keypoint was obtained by first computing the gradient magnitudes and orientations of pixels in the neighborhood region of the keypoint, using the scale of the keypoint to select a proper Gaussian kernel to blur the image. In order to achieve orientation invariance, the coordinates of the descriptor and the gradient orientations were rotated relative to the keypoint orientation. The orientation histograms within the subregions around the keypoint were computed and combined into the SIFT feature vector. This vector was normalized to improve the invariance to changes of illumination. Gradient Location and Orientation Histogram (GLOH) [17] extended SIFT by allowing SIFT descriptor to be computed on a log-polar location grid.

HOG. Histogram of Oriented Gradients proposed by Dalal and Triggs [18] was similar to SIFT. It computed the histograms of gradient orientations in different subregions. Different from SIFT, which was computed on detected sparse keypoints, HOG was sampled from a dense and uniform grid and was improved by local contrast normalization in overlapping spatial blocks. Integral Histogram of Oriented Gradients (IHO) [19] is an approximation of HOG and can be efficiently computed using integral images.

MSER. Instead of detecting keypoints, Maximally Stable Extremal Regions (MSER) proposed by Matas et al. [20] detected regions which were darker or brighter than surroundings. It was affinely-invariant and robust to changes of illuminations. It was extended to colour in [21].

SURF. Bay et al. [22] proposed the SURF (Speeded Up Robust Features) descriptor, which could be efficiently computed using integral images. The neighborhood of a pixel was uniformly adapted into $P \times Q$ spatial bins. The SURF descriptor was calculated by accumulating the sum of Haar wavelet responses at different spatial bins. Let d_x and d_y be the Haar wavelet responses in the horizontal and vertical directions. The descriptor has a four-dimensional vector $(\sum d_x, \sum \|d_x\|, \sum d_y, \sum \|d_y\|)$ for each spatial bin. The resulting $4 \times P \times Q$ dimensions SURF descriptor was L1-normalized.

Spin image and RIFT. Lazebnik et al. [23] proposed two rotation-invariant descriptors, spin image and RIFT (Rotation-Invariant Feature Transform). The spin image was a two-dimensional histogram of image intensities and their distance to the keypoint. To construct the RIFT descriptor, the circular normalized patch around the keypoint was divided into concentric rings of equal width and a gradient orientation histogram was computed within each ring.

Most descriptors described above were applied to intensity images. To increase illumination invariance and discriminative power, color descriptors were proposed. An evaluation of different color descriptors can be found in [24]. It was shown that the combination of different filter-banks of visual descriptors could improve the performance [25].

These filter bank responses and invariant descriptors can be computed at image patches densely sampled or at sparse interest points. Results in [26] showed that dense sampling improved the performance because it captured the most information, but its computation was expensive.

3.2.2 Textons and Visual Words

In semantic object segmentation, the responses of filter-banks or visual descriptors are usually further quantized to textons or visual words[1] according to a learned codebook. Since the histograms of textons or visual words will be used as the input of classifiers at later stages, the design of codebooks should consider both distinctiveness and repeatability. This means that it should try to assign image patches of different object classes to different codewords and to assign image patches of the same object class to the same codeword. The codebook should be compact in order to avoid overfitting of the classifiers at later stages. Because there are a huge number of image patches in data collections, memory and computation efficiency is another issue to be considered when learning the codebooks.

K-means is the most commonly used clustering methods to generate the codebooks. Some examples of visual words obtained by k-means are shown in Fig. 3.5. Since the distribution of image patches in the filter-bank space or in the descriptor space is far from uniform, one of the disadvantages of k-means is that it clusters centres almost exclusively around the densest few regions in descriptor space and cannot over other informative regions well. Based on this consideration, Jurie et al. [27] proposed a new approach building codebooks using mean shift. Some patches in the dense regions were removed and the learned codebooks were more informative.

K-means has high computational cost. It also has the difficulty of balancing the distinctiveness and repeatability by choosing different sizes of codebooks. If the size of the codebook is too small, image patches of different object classes will fall into the same bin. At the other extreme, image patches around the same keypoint observed in different images will fall into different bins. To overcome

---

[1] Textons are quantized responses of filter-banks and visual words are quantized visual descriptors.

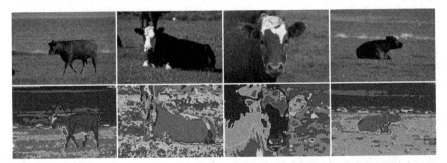

Fig. 3.5 Examples of visual words obtained by the filter-banks proposed in [2] and k-means. The first row are images and the second row are visual words. Colors represent different visual words

these difficulties, Nister et al. [28] proposed the vocabulary tree constructed by hierarchical k-means. It allowed a larger and more discriminatory codebook to be used efficiently. Moosmann et al. [29] proposed Extremely Randomized Clustering Forests, which were ensembles of randomly created clustering trees, to learn the codebook. It provided more accurate results and was faster than k-means. Elkan [30] used the triangle inequality to dramatically accelerate k-means, while guaranteed always computing exactly the same result as the standard k-means.

K-means assumed hard assignment, i.e. exactly assigning a single visual word to one image feature. If an image feature is relevant to multiple textons or visual words, only the best is selected. If none of the codewords in the codebook well represent the image feature, the best one is still assigned to the image feature. These may cause problems during object segmentation. van Gemert et al. [31] created codebooks using kernel density estimation. It modeled the uncertainty between visual words and image features.

The above approaches are unsupervised. Some supervised approaches learned codebooks incorporate semantic information. These codebooks were more compact and discriminative. Winn et al. [2] learned an optimally compact visual codebook by pairwise merging of visual words given segmented images for training. Shotton et al. [32] proposed semantic texton forests, which were randomized decision forests [33] and were learned from image pixels. Perronnin et al. [34] learned different codebooks for different object classes by adapting a universal codebook, which described the content of all the classes of images, using class-specific data. Both the universal codebook and adapted class-codebooks were used for classification.

3.3 Object Segmentation Using Discriminative Approaches

3.3.1 Classifiers on Local Appearance

The obtained histograms of textons or visual words within local regions capture the features of local appearance and are usually used as the input of classifiers to predict object labels. Support Vector Machines (SVM) and Boosting are widely used to model the appearance of object classes. Marsalek and Schmid [35] estimated the shape mask of an object and its object category using nonlinear SVM and with χ^2 distance. The appearance of the object within the shape mask was represented by a histogram of visual words. Shotton et al. [32] used the texton histograms and region priors, which were calculated from their proposed semantic texton forests, of image regions as input of a one-vs-others SVM classifier to assign image regions into different object classes. Gould et al. [36] used the boosting classifier to predict the label of each pixel. Tahir et al. [25] used Spectral Regression Kernel Discriminant Analysis(SRKDA) [37] and achieved better results than SVM on PASCAL VOC 2008 [6]. It was also much more efficient than Kernel Discriminant Analysis (KDA). Aldavert et al. [38] proposed an integral linear classifier, which used integral images to efficiently calculate the outputs of linear classifiers based on histograms of visual words at the pixel level.

3.3.2 Conditional Random Fields

Although classifiers such as SVM and Boosting can predict the object label of a pixel based on the appearance within its neighborhood, they cannot capture local consistency of other contextual features, such as "sky" appears above buildings but not the other way around. Local appearance, local consistency and contextual features can be well incorporated under a Conditional Random Fields (CRF) framework.

3.3.2.1 Multiscale Conditional Random Fields

He et al. [39] were the first to use CRF for semantic object segmentation. Their proposed CRF framework is described as following. Suppose $\mathbf{X} = \{x_i\}$ are image patches and $\mathbf{Z} = \{z_i\}$ are their object class labels. In [39], the conditional distribution over $\mathbf{Z}$ given by input $\mathbf{X}$ was defined by multiplicatively combining component conditional distributions.

$$P(\mathbf{Z}|\mathbf{X}) \propto P_C(\mathbf{Z}|\mathbf{X})P_R(\mathbf{Z}|\mathbf{X})P_G(\mathbf{Z}|\mathbf{X}). \tag{3.1}$$

P_C, P_R, and P_G capture statistical structures at three different spatial scales: local classifier, regional features, and global features (see Fig. 3.6).

The local classifier P_C produces a distribution over the label z_i given by its image patch x_i as input,

$$P_C(\mathbf{Z}|\mathbf{X},\lambda) = \prod_i P_C(z_i|x_i,\lambda), \tag{3.2}$$

where λ is the parameter of the local classifier. A 3-layer multilayer perceptron (MLP) was used in [39].

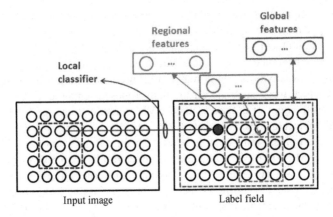

Fig. 3.6 A graphical representation of CRF. Reproduced from [39]

The regional features P_R represent local geometric relationships between objects. They avoid impossible combinations of neighboring objects such as "ground is above sky" and also encourage the segmentation results to be spatially smooth. A collection of regional features are learned from the training data. Let r be the index of regions and a be the index of the different regional features within each region, and j be the index of image patches within in region r. P_R is defined as

$$P_R(\mathbf{Z}, \mathbf{f}) \propto \exp\left\{ \sum_{r,a} f_{r,a} \mathbf{w}_a^T \mathbf{z}_r \right\}. \tag{3.3}$$

$\mathbf{f} = \{f_{r,a}\}$ are binary hidden regional variables. $f_{r,a} = 0, 1$ indicating the feature a in region r exists or not. $\mathbf{w}_a = [w_{a,1}, \ldots, w_{a,J}, \alpha_a]$ are parameters and α_a is a bias term. $w_{a,j}$ connects $f_{r,a}$ with $z_{r,j}$ and specifies preferences for the possible label value of $z_{r,j}$. $\mathbf{z}_r = [z_{r,1}, \ldots, z_{r,J}, 1]$. P_R is high of $\mathbf{z}_r$ matches $\mathbf{w}_a$ and $f_{r,a} = 1$ or $\mathbf{z}_r$ does not match $\mathbf{w}_a$ and $f_{r,a} = 0$.

The global feature P_G is defined over the whole image,

$$P_G(\mathbf{Z}, \mathbf{g}) \propto \exp\left\{ \sum_b g_b \mathbf{u}_b^T \mathbf{Z} \right\}. \tag{3.4}$$

b is the index of the global label patterns, which are encoded in the parameters $\{\mathbf{u}_b\}$. $\mathbf{g} = \{g_b\}$ are the binary hidden global variables.

Both hidden variables $\mathbf{f}$ and $\mathbf{g}$ can be marginalized, leading to

$$P_R(\mathbf{Z}) \propto_{r,a} \left[1 + \exp\left(\mathbf{w}_a^T \mathbf{z}_r \right) \right], \tag{3.5}$$

$$P_G(\mathbf{Z}) \propto_b \left[1 + \exp\left(\mathbf{u}_b^T \mathbf{Z} \right) \right]. \tag{3.6}$$

Thus (3.2) has a closed form,

$$P(\mathbf{Z}|\mathbf{X};\theta) \propto \prod_i P_C(z_i|x_i,\lambda) \times \prod_{r,a} \left[1 + \exp\left(\mathbf{w}_a^T \mathbf{z}_r\right)\right] \times \prod_b \left[1 + \exp\left(\mathbf{u}_b^T \mathbf{Z}\right)\right]. \quad (3.7)$$

$\theta = \{\lambda, \{\mathbf{w}_a\}, \{\mathbf{u}_b\}\}$ are parameters. They are learned from a training by maximizing the conditional likelihood in [39]. Once the parameters are learned, the object class labels are inferred by maximizing posterior marginals.

3.3.2.2 TextonBoost

Under the CRF framework, Shotton et al. [40] proposed TextonBoost to learn a discriminative model of object classes incorporating texture, layout, and context information. Their CRF includes four types of potentials: texture-layout, color, location, and edge.

$$\log P(Z|X,\theta) = \sum_i \overbrace{\psi_i(z_i,\mathbf{X};\theta_\psi)}^{\text{texture−layout}} + \overbrace{\pi(c_i,x_i;\theta_\pi)}^{\text{color}} + \overbrace{\ell(z_i,i;\theta_\ell)}^{\text{location}}$$

$$+ \sum_{(i,j)\in\varepsilon} \overbrace{\xi(z_i,z_j,g_{ij}(\mathbf{X});\theta_\xi)}^{\text{edge}} - \log C(\theta,\mathbf{X}), \quad (3.8)$$

where i and j are indices of pixels, $(i,j) \in \varepsilon$ are two neighboring pixels, $\theta = \{\theta_\psi, \theta_\pi, \theta_\ell, \theta_\xi\}$ are parameters, and $C(\theta, \mathbf{X})$ is a normalization term.

The texture-layout potentials are provided by a boosting classifier combining a set of discriminative features called texture-layout filters. The neighborhood of pixel i is partitioned into regions by a predefined spatial kernel. Each texture-layout $v_{[r,t]}(i)$ is the number of pixels with texton t in region r. Therefore, texture-layout filters are histograms of textons over defined spatial kernels. They capture texture, spatial layout, and textural context. Discriminative texture-layout filters are selected as weak classifiers and combined into a powerful classifier by Joint Boost [41]. Joint Boost allows to share weak classifiers among different object classes and the learn classifier has better generalization.

The color potentials model the color distribution of each object class using Gaussian mixture models in CIELab color space.

The location potentials model the dependence between the locations of pixels and object classes. For example, trees and sky tend to appear in the top regions of images while roads tend to appear in the bottom regions of images.

In the edge potentials, g_{ij} measures the edge features between neighbor pixels. A penalty is added if two neighboring pixels have different object class labels unless there is a strong edge between them.

TextonBoost was evaluated on 21 object classes from the MSRC database and achieved 72.2% overall accuracy [40]. The confusion matrix is shown in Fig. 3.7.

True class / inferred class	building	grass	tree	cow	sheep	sky	aeroplane	water	face	car	bike	flower	sign	bird	book	chair	road	cat	dog	body	boat
building	61.6	4.7	9.7	0.3		2.5	0.6	1.3	2.0	2.6	2.1		0.6	0.2	4.8		6.3	0.4		0.5	
grass	0.3	97.6	0.5										0.1							1.3	
tree	1.2	4.4	86.3	0.5		2.9	1.4	1.9	0.8	0.1							0.1		0.2	0.1	
cow		30.9	0.7	58.3				0.9	0.4			0.4			4.2					4.1	
sheep	16.5	25.5	4.8	1.9	50.4										0.6		0.2				
sky	3.4	0.2	1.1			82.6		7.5									5.2				
aeroplane	21.5	7.2				3.0	59.6	8.5													
water	8.7	7.5	1.5	0.2		4.5		52.9		0.7	4.9			0.2	4.2		14.1	0.4			
face	4.1		1.1						73.5	7.1					8.4			0.4	0.2	5.2	
car	10.1		1.7							62.5	3.8		5.9	0.2			15.7				
bike	9.3		1.3							1.0	74.5		2.5			3.9	5.9		1.6		
flower		6.6	19.3	3.0								62.8			7.3		1.0				
sign	31.5	0.2	11.5	2.1		0.5		6.0		1.5		2.5	35.1		3.6	2.7	0.8	0.3		1.8	
bird	16.9	18.4	9.8	6.3	8.9	1.8		9.4						19.4			4.6	4.5			
book	2.6		0.6							0.4		2.0			91.9					2.4	
chair	20.6	24.8	9.6	18.2		0.2					3.7				1.9	15.4	4.5		1.1		
road	5.0	1.1	0.7					3.4	0.3	0.7	0.6		0.1	0.1	1.1		86.0			0.7	
cat	5.0		1.1	8.9				0.2		2.0					0.6			28.4	53.6	0.2	
dog	29.0	2.2	12.9	7.1				9.7										8.1	11.7	19.2	
body	4.6	2.8	2.0	2.1	1.3	0.2				6.0	1.1						9.9	1.7	4.0	2.1	62.1
boat	25.1		11.5			3.8		30.6		2.0	8.6		6.4	5.1			0.3				6.6

Fig. 3.7 Confusion matrix of object segmentation by TextonBoost [40] on the MSRC 21 database. The figure is reproduced from [40]

The experimental evaluation showed that although the texture-layout potentials had the most significant contribution to semantic object segmentation, CRF significantly improved the accuracy of results.

3.3.2.3 Other Approaches Based on Conditional Random Fields

Other semantic object segmentation approaches based CRF were proposed. Fulkerson et al. [42] treated superpixels [43], which were small regions obtained from a conservative oversegmentation, as basic units of segmentation. They assumed that superpixels allowed to measure histograms of visual words on a natural adaptive domain rather than on a fixed patch window. Moreover, superpixels tended to preserve boundaries and created more accurate segmentation. A one-vs-others SVM classifier with a RBF-χ^2 kernel was constructed on the histograms of visual words found in each superpixel. This local classifier was used in a CRF operating on the superpixel graph. CRF was used to add spatial regularization by requiring that if two neighboring superpixels share a long boundary and were similar in appearance, they tended to have the same class label. It discouraged small isolated regions and reduced misclassifications that occurred near the edges of objects. He et al. [44] also first oversegmented images into superpixels. Superpixels were labeled under a mixture of CRF. Images in a database were grouped into several contexts and each context was modeled by a separate CRF.

Torralba et al. [45] proposed Boosted Random Fields for object detection and segmentation. Boosting was used to learn the graph structure and local evidence of a conditional random field. The graph structure of CRF was learned using boosting to select from a dictionary connectivity templates, which were derived from labeled segmentations. It exploited the contextual correlations between object classes. Rabinovich et al. [46] explicitly defined the interactions between object classes as semantic context and incorporated it into CRF. The semantic context was modeled as the co-occurrence of object labels and was learned both from the training data and Google Sets.[2]

Quattoni et al. [47] used CRF for part-based object recognition and detection. CRF was used to model the spatial arranges of object parts. Ma and Grimson [48] proposed a coupled CRF to decompose the images into contour and texture and to model their interaction. The decomposed low-level cues were adaptively combined for object recognition and different discriminative cues for different object classes were fully leveraged. Reynolds and Murphy [49] proposed a tree-structured CRF for object segmentation.

3.4   Object Segmentation Using Topic Models

The discriminative approaches described above required training data to be labeled at pixel-level. If there are a large number of object classes to be modeled, the labeling work is very expensive. Some researchers started to explore approaches of learning the models of object classes from a collection of images or videos without supervision or with weak supervision (such as using training data labeled at image-level). Inspired by the success of topic models, such as Probabilistic Latent Semantic Analysis (pLSA) [9] and Latent Dirichlet Allocation (LDA) [10], in the applications of language processing, they have been also applied to semantic object segmentation in recent years. Under pLSA or LDA, words, such as "professor" and "university", often co-existing in the same documents are clustered into the same topic, such as "education". The models of topics are automatically without supervision. The word-document analysis has been applied to object segmentation through mapping the concepts of "words" and "documents" to the image and video domains. For example, if images are treated as documents and visual words (or textons) are treated as words, with the assumption that visual words of the same object classes often co-exist in the same images, the models of object classes can be learned as the models of topics. Object classes are treated as topics. Since an image may include objects of several classes, it is modeled as a mixture of topics. An advantage of such an approach is that manually segmenting objects at the pixel level is not required for training. Some proposed approaches [11, 50, 51] were totally unsupervised. Some required labeling at the image level [52, 53]. Some semantic object segmentation approaches based on topics models will be reviewed in this section.

---

[2] http://labs.google.com/sets

3.4.1 pLSA and LDA

Sivic et al. [50] discovered the object classes from a set of unlabeled images and segmented images into different object classes using pLSA and LDA. They modeled an image as a bag of visual words and ignored any spatial relationships among visual words. Suppose there are M images in the data set. Each image j has N_j visual words. Each visual word w_{ji} is assigned one of the K object classes according to its label z_{ji}. Under pLSA, the joint probability $P(\{w_{ji}\}, \{d_j\}, \{z_{ji}\})$ has the form of the graphical model shown in Fig. 3.8a. The conditional probability $P(w_{ji}|d_j)$ marginalizing over topics z_{ji} is given by

$$P(w_{ji}|d_j) = \sum_{k=1}^{K} P(z_{ji} = k|d_j)P(w_{ji}|z_{ji} = k). \qquad (3.9)$$

$P(z_{ji} = k|d_j)$ is the probability of object class k occurring in image d_j. $P(w_{ji}|z_{ji} = k)$ is the probability of visual word w_{ji} occurring in object class k and is the model of object class k. Fitting the pLSA model involves determining $P(w_{ji}|z_{ji})$ and $P(z_{ji} = k|d_j)$ by maximizing the following objective function using the Expectation Maximization (EM) algorithm:

$$L = \prod_{j=1}^{M} \prod_{i=1}^{N_j} P(w_{ji}|d_j). \qquad (3.10)$$

Images are segmented into objects with semantic meanings based on the labels z_{ji} of visual words.

pLSA is a generative model only for training images but not for new images. This shortcoming has been addressed by LDA, whose graphical model is shown in Fig. 3.8b. Under LDA, $\{\phi_k\}$ are models of object classes and are discrete distributions over the codebook of visual words. They are generated from a Dirichlet prior $Dir(\phi_k; \beta)$ given by β. Each image j has a multinomial distribution π_j over K object

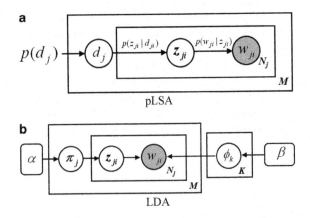

Fig. 3.8 Graphical models of pLSA and LDA

Table 3.1 Confusion table of using pLSA for image classification on a data set of five object categories from the Caltech 101 database [55]. Class number is equal to 7 in pLSA. Three classes correspond to the background. The result was reported in [50]

True class →	Faces	Motorbikes	Airplanes	Cars	Background
Class 1 - Faces	94.02	0.00	0.38	0.00	1.00
Class 2 - Motorbikes	0.00	83.62	0.12	0.00	1.25
Class 3 - Airplanes	0.00	0.50	95.25	0.52	0.50
Class 4 - Cars	0.46	0.88	0.38	98.1	3.75
Class 5 - Background I	1.84	0.38	0.88	0.26	41.75
Class 6 - Background II	3.68	12.88	0.88	0.00	23.00
Class 7 - Background III	0.00	1.75	2.12	1.13	28.75

classes and it is generated from a Dirichlet prior $Dir(\pi_j; \alpha)$. Each patch i on image j is assigned to one of the K object classes and its label z_{ji} is sampled from a discrete distribution $Discrete(z_{ji}; \pi_j)$ given by π_j. The observed visual word w_{ji} is sampled from the model of its object class: $Discrete(w_{ji}|\phi_{z_{ji}})$. α and β are hyperparameters. ϕ_k, π_j and z_{ji} are hidden variables to be inferred. The inference can by implemented by variational methods [10] or collapsed Gibbs sampling [54]. Under LDA, if two visual words often co-occur in the same images, one of the object class models will have large distributions on both of them. pLSA and LDA perform similarly on image classification and object segmentation and their results were promising especially when each image only contained one object. As reported by [50], on a data set consisting of 4,090 images of five categories from the Caltech 101 database [55], the image classification accuracy achieved by pLSA was 92.5% (see Table 3.1) and its object segmentation accuracy was 49%. Both pLSA and LDA requires the number of object classes to be known in advance. As an extension, Hierarchical Dirichlet Process (HDP) proposed by Teh et al. [54] could automatically learn the number of object classes from data using Dirichlet Processes [56] as priors.

3.4.2 SLDA

A shortcoming of using pLSA and LDA to segment objects is to treat an image as a document of visual words ignoring the spatial structure among visual words. The assumption that if two types of patches are from the same object class, they often appear in the same images is not strong enough. As an example shown in Fig. 3.9, although the sky is far from the vehicles, if they often exist in the same images in the data set, they would be clustered into the same topic (object class) by pLSA or LDA. Since most parts of this image are sky and building, an image patch on a vehicle is likely to be labeled as building or sky as well. Such problems can be solved if the document of an image patch, such as the yellow patch in Fig. 3.9, only includes patches falling within its neighborhood, marked by the red dashed window in Fig. 3.9 instead of the whole image.

Fig. 3.9 There will be some problems (see text) if the whole image is treated as one document when using LDA to discover classes of objects

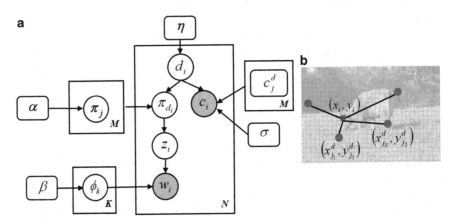

Fig. 3.10 (a) Graphical model of SLDA. (b) Add spatial information when designing documents. Each document is associated with a point (marked in magenta color). These points are densely placed over the image. If an image patch is close to a document, it has a high probability to be assigned to that document

With the assumption that if two types of image patches are from the same object class, they are not only often in the same images but also close in space, a Spatial Latent Dirichlet Allocation (SLDA) was proposed in [11]. Under SLDA, the word-document assignment becomes a hidden random variable. There is a generative procedure to assign words to documents. When visual words are close in space or time, they have a high probability to be grouped into the same document. The graphical model SLDA is shown in Fig. 3.10. The N visual words in an image set are assigned to M documents. d_j is a hidden variable indicating the document assignment of visual word i. Each document j is associated with a hyperparameter $c_j^d = (g_j^d, x_j^d, y_j^d)$, where g_j^d is the index of the image where the document is placed and (x_j^d, y_j^d) is the location of the document. Besides the word value w_{ji}, the location (x_i, y_i) and image index g_i of a word i are observed and stored in variable $c_i = (g_i, x_i, y_i)$. The generative procedure is as following:

1. For a topic k, a multinomial parameter ϕ_k is sampled from Dirichlet prior $\phi_k \sim Dir(\beta)$.
2. For a document j, a multinomial parameter π_j over the K topics is sampled from Dirichlet prior $\pi_j \sim Dir(\alpha)$.
3. For a word (image patch) i, a random variable d_i is sampled from prior $p(d_i|\eta)$ indicating to which document word i is assigned. We choose $p(d_i|\eta)$ as a uniform prior.
4. The image index and location of word i are sampled from distribution $p(c_i|c_{d_i}^d, \sigma)$. We may choose this as a Gaussian kernel.

$$
p\left((g_i, x_i, y_i) \,\middle|\, \left(g_{d_i}^d, x_{d_i}^d, y_{d_i}^d\right), \sigma\right) \propto \delta_{g_{d_i}^d}(g_i) \exp\left\{ -\frac{\left(x_{d_i}^d - x_i\right)^2 + \left(y_{d_i}^d - y_i\right)^2}{\sigma^2} \right\},
$$

$p(c_i|c_{d_i}^d, \sigma) = 0$ if the word and the document are not in the same image.
5. The topic label z_i of word i is sampled from the discrete distribution of document d_i, $z_i \sim Discrete(\pi_{d_i})$.
6. The value w_i of word i is sampled from the discrete distribution of topic z_i, $w_i \sim Discrete(\phi_{z_i})$.

In [11] both LDA and SLDA were evaluated on the MSRC data set [2] with 240 images for object segmentation. The detection rate and false alarm rate of four classes (cows, cars, faces, and bicycles) are shown in Table 3.2. Some examples are shown in Fig. 3.11. The segmentation results of LDA were noisy since spatial information was not considered. The patches in the same image were likely to have the same labels. SLDA achieved better results.

In [11] SLDA was also used to segment objects from a video sequence. All the frames were treated as a collection of images and their temporal order was ignored. Figure 3.12 shows results on two sampled frames. LDA could not segment out any objects. SLDA clustered image patches into tigers, rock, water, and grass.

Table 3.2 Detection(Det) rate and False Alarm (FA) rate of LDA and SLDA on MSRC [2]. The results are from [11]

	Cows		Cars		Faces		Bicycles	
	Det rate	FA rate	Det rate	FA rate	Det rate	FA rate	Det rate	FA rate
LDA	0.3755	0.5576	0.5552	0.3963	0.7172	0.5862	0.5563	0.5285
SLDA	0.5662	0.0334	0.6838	0.2437	0.6973	0.3714	0.5661	0.4217

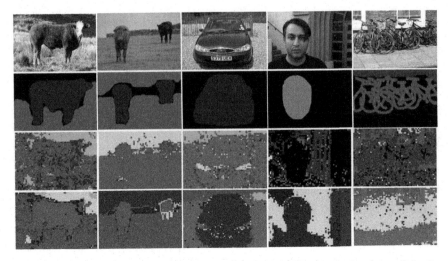

Fig. 3.11 Examples of object segmentation results by LDA and SLDA. The images are from the MSRC data set [2]. The first row shows example images. The second row uses manual segmentation and labeling as ground truth. The third row is the LDA result and the fourth row is the SLDA result. Under the same labeling approach, image patches marked in the same color are in one object cluster, but the meaning of colors changes across different labeling methods. The results are from [11]

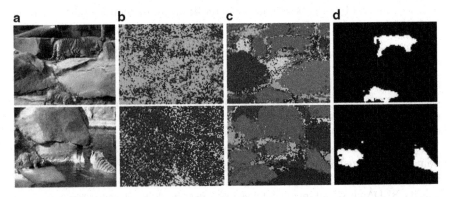

Fig. 3.12 Object segmentation from a video sequence. The first column shows two frames in the video sequence. In the second column, the patches in the two frames as are labeled as different object classes using LDA. The third column plots the object class labels using SLDA. In the fourth column, tigers are segmented out by choosing all the patches of the class marked by red color. The results are from [11]

3.4.3 Other Topic Models of Including Spatial Information

Some other topic models were also proposed to include spatial information. Russell et al. [51] first obtained multiple segmentations of each image at different scales using normalized cut [57] and then treated each segment instead of an image as a document. These segments captured the spatial relationships among visual words. Some good segments are sifted from bad ones for each discovered object class.

Verbeek et al. [52] proposed two aspect-based spatial field models by combining pLSA/LDA with Markov Random Fields (MRF). One is based on averaging over forests of minimal spanning trees linking neighboring image regions. A tree-structure prior is imposed to the object class labels $Z_j = \{z_{ji}\}$ of image patches in image j,

$$P(Z_j) \propto \exp\left(\sum_i \psi(z_{ji}, z_{j\chi(i)}) + \log \theta_j \right), \qquad (3.11)$$

where $\chi(i)$ is the unique parent of patch i in the tree, and $\psi(z_{ji}, z_{j\chi(i)})$ is a pair-wise potential,

$$\psi(z_{ji}, z_{j\chi(i)}) = \rho[z_{ji} = z_{j\chi(i)}]. \qquad (3.12)$$

The other model applies an efficient chain-based Expectation Propagation method for regular 8-neighbor Markov Random Fields. The prior over Z_j is given by

$$P(Z_j) \propto \exp\left(\sum_{i \sim i'} \psi(z_{ji}, z_{ji'}) + \log \theta_j \right), \qquad (3.13)$$

where $i \sim i'$ enumerates spatial neighbor patches i, i' in image j. MRF captures the local spatial dependence of image patches. These two models were trained using either patch-level labels or image-level labels. Tested on 240 images of nine object categories from the MSRC data set, when trained using patch-level labels, they achieved object segmentation accuracy of 80.2% and when trained using image-level labels, the accuracy of 78.1% was achieved. The accuracies of pLSA were 78.5% and 74.0% respectively under these two settings. The similar idea was also explored in [58] and a Dirichlet process mixture was introduced to automatically learn the number of object classes from data. This framework was extended to Conditional Random Field (CRF) [4] to integrated both local and global features in the images [53, 59].

Sudderth et al. [60] proposed a Transformed Dirichlet Process (TDP) model to jointly solve the problem of scene classification and object segmentation. This approach coupled topic models with spatial transformations and consistently accounted for geometric constraints. The spatial relationships of different parts of objects were explicitly modeled under a hierarchical Bayesian model. Cao et al. [61] proposed a Spatially Coherent Latent Topic Model (Spatial-LTM) to simultaneously classify scene categories and segment objects. It oversegmented images into regions of coherent latent topic model and coherent latent topic model was considered as visual words. It enforced the spatial coherency of the model by requiring that only one single latent-topic was assigned to the image patches within each region.

3.5 Object Segmentation in Videos

A video is composed of a sequence of images. Different from still image segmentation, video segmentation should take account the temporal information. Many statistical models have been proposed for video segmentation, either generative or discriminative. In the discriminative model, a large number of expensive labeled data is required to train an excellent classifier. On the contrary, the generative model can handle the incomplete data problem and address the large number of unlabeled data via small number of expensive labeled data. Therefore, the generative model is popular for video segmentation. On the other hand, the discriminative model relaxes the conditional independence assumption and has better predictive performance than the generative model. This attracts many attentions to the discriminative model in video segmentation. MRFs [62, 63] and CRFs [64–67] are representative generative and discriminative models in video segmentation, respectively.

Let $\mathbf{X} = \{x_i\}_{i \in S}$ and $\mathbf{Z} = \{z_i\}_{i \in S}$ be the observation and labels of a video, where $S = \{s_i\}$ is the set of units (they can be pixels, patches, or semantic regions) in the video. Then video segmentation is to maximize the posterior $p(\mathbf{Z}|\mathbf{X})$.

3.5.1 MRF Model

In the MRF model, the posterior is expressed proportioned to the joint probability using the Baye's rule as:

$$p(\mathbf{Z}|\mathbf{X}) \propto p(\mathbf{Z}|\mathbf{X}) = p(\mathbf{X}|\mathbf{Z})p(\mathbf{Z}), \tag{3.14}$$

where the prior $p(\mathbf{Z})$ is modeled as a MRF.

In the MRF model, the strong assumption of conditional independency of the observed data is enforced. Therefore, the likelihood $p(\mathbf{X}|\mathbf{Z})$ is assumed to have a factorized form, i.e.,

$$p(\mathbf{X}|\mathbf{Z}) = \prod_{s_i \in S} p(x_i|z_i). \tag{3.15}$$

Here $p(x_i|z_i)$ indicates the probability that the unit s_i has the label z_i based on the observation x_i at s_i. Here x_i can be features incorporating the color, texture, and motion information. To adapt to changes of environment, some features robust to illumination changes are utilized, like gradient direction, shadow models, and color co-occurrence.

To model the distribution of $p(x_i|z_i)$, several ways have been proposed. The most traditional approach is model the distribution in terms of the Gaussian Mixture Models (GMMs) and the Expectation Maximization (EM) algorithm is used to estimate the model parameters. The GMM model has several shortcomings: it is sensitive to the initialization, the EM algorithm takes long time to converge, and a suitable number of Gaussian components have to be set. To address these problems, a non-parametric way, smoothed histograms in the YUV color space [64], has been proposed. It learns the histograms from some labeled region and stored in 3D look-up tables with smoothing. Then the value of $p(x_i|z_i)$ is searched from the histogram tables.

In the MRF model, $p(\mathbf{Z})$ is used to enforce the Markov properties of the labels. In the Bayesian view, the prior $p(\mathbf{Z})$ does not depend on the observed data $\mathbf{X}$. It is assumed to be an Potts model, i.e.,

$$p(\mathbf{Z}) = \exp\left(\sum_{s_i \in S}\sum_{s_j \in \mathcal{N}_i} \lambda T(z_i \neq z_j)\right), \tag{3.16}$$

where $\mathcal{N}_i$ is the neighborhood system of s_i, λ is a negative constant, and $T(\cdot) = 1$ if its argument is true and $T(\cdot) = 0$ if false. In video segmentation, the neighborhood system includes two parts, the spatial and temporal neighborhoods. The prior in the spatial neighborhood system incorporates the spatial smoothness constraint, which can reduce the effect of noise. The prior in the temporal neighborhood system is used to incorporate the inter-frame information. In the case of binary class problem (e.g., in foreground/background segmentation, $z_i \in \{1, -1\}$), the prior $p(\mathbf{Z})$ can be transformed as an isotropic Ising model, i.e.,

$$p(\mathbf{Z}) = \exp\left(\sum_{s_i \in S}\sum_{s_j \in \mathcal{N}_i} \lambda z_i z_j\right). \tag{3.17}$$

As noted above, the prior $p(\mathbf{X})$ does not depend on the observed data. But in the applications of video segmentation, observed data-dependent prior is necessary. In the part of spatial neighborhood system, the contrast information is incorporated by modulating the prior according to the intensity gradients. In the temporal part, the intensity difference is used to control the probability of s_i and s_j having the same label. Therefore, in video segmentation, the prior is expressed as

$$p(\mathbf{Z}) = \exp\left(\sum_{s_i \in S}\sum_{s_j \in \mathcal{N}_i} \lambda T(z_i \neq z_j) \cdot \exp\left(-\Delta_{i,j}^2/\sigma\right)\right), \tag{3.18}$$

where $\Delta_{i,j}$ is the intensity difference between s_i and s_j and σ is a positive constant. From the equation we can see that if s_i and s_j have a larger intensity difference, then they have a higher probability of being different labels.

Combining (3.14), (3.15), and (3.18), the posterior in MRF model is expressed as

$$p(\mathbf{X}|\mathbf{Z}) = \frac{1}{C}\exp\left(\sum_{s_i \in S} \log\left(p(x_i|z_i)\right) + \sum_{s_i \in S}\sum_{s_j \in \mathcal{N}_i} \lambda_m T(z_i \neq z_j)\right), \tag{3.19}$$

where C is the partition function and $\lambda_m = \lambda \exp(-\Delta_{i,j}^2/\sigma)$.

3.5.2 CRF Model

Compared with MRFs, the CRF model formulates the posterior $p(\mathbf{Z}|\mathbf{X})$ directly instead of formulating the joint probability $p(\mathbf{Z},\mathbf{X})$ via the likelihood $p(\mathbf{X}|\mathbf{Z})$ and $p(\mathbf{Z})$ by the Baye's rule. Generally, the posterior in MRFs is written as,

$$p(\mathbf{Z}|\mathbf{X}) = \frac{1}{C}\exp\left(\sum_{s_i \in S} u_i(z_i,\mathbf{X}) + \sum_{s_i \in S}\sum_{s_j \in \mathcal{N}_i} v_{ij}(z_i,z_j,\mathbf{X})\right), \qquad (3.20)$$

where C is the partition function, $-u_i$ and $-v_{ij}$ are the unary and pairwise potential, respectively.

Comparing (3.20) with (3.19), the definitions of unary potential and the pairwise potential are different between MRFs and CRFs. In CRFs, the unary potential is a function in the term of the whole observed data $\mathbf{X}$, while in MRFs the unary potential for s_i is a function in term of observed data at s_i due to the conditionally independent assumption. Theoretically, in MRFs, the pairwise potential is a function of only labels (actually a function of labels and the intensity difference in the applications of video segmentation) while it is a function of labels and the whole observed data $\mathbf{X}$ in CRFs.

Since the potentials are in term of the whole observed data in CRFs, they are designed by using some arbitrary local discriminative classifiers. In discriminative classifiers, it is important to select a good feature space. Compared with MRFs, the CRF model selects more discriminative features besides colors, constant, and other features used in MRFs. For example in [65], texture, location, and histogram of oriented gradient (HOG) features are used for scene labeling. In [66], motion-shape cues are used for bilayer video segmentation. The features of "motons" (related to textons) are used for modeling the motion information in videos. A shape-filter modeling long-range correlation is selected to describe the shape features. Actually, any fusion of discriminative features used in images can be selected in video segmentation. The difference between the video and image applications is good discriminative features describing the motion information, which may be used to improve the video segmentation results.

The second important thing in the discriminative model is classifier selection. In common, the classification algorithms build strong classifiers from a combination of weak classifiers. The difference between these algorithms is the way that the weak classifiers combine. In [66], the authors construct a tree cube taxonomy for helping to select classification algorithms. Figure 3.13 is the tree cube taxonomy of classifiers. The origin is the weak learner and the axes H, A, and B are three basic ways of combining weak learners: hierarchically (H), by averaging (A), and via boosting (B). Different strong classifiers, i.e., different combinations of weak classifiers, correspond to different paths along the edges of the cube in Fig. 3.13.

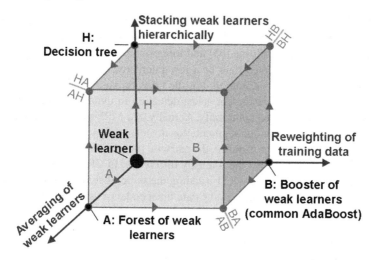

Fig. 3.13 The tree cube taxonomy of classifiers. The figure is taken from [66]

3.5.3 MRFs Vs. CRFs

This section summarizes some main differences between MRFs and CRFs.

Formulation: In MRFs, the posterior is proportioned to the joint probability using the Baye's rule, and the joint probability is modeled by defining the likelihood and prior while CRFs model the posterior directly. In MRFs, the unary and pairwise potentials are functions of observed data at individual site and only the labels, respectively. While in CRFs, the unary and pairwise potentials are functions of the whole observed data and labels.

Feature space: In MRFs, since the distributions of the observed data should be modeled, low-dimension features, like color and motion, are used in common. While in CRFs, more complex discriminative features would be selected to improve the predictive performance.

Performance: Compared with CRFs, MRFs can handle data missing problem and new class adding problem. While CRFs have better predictive performance since CRFs model the posterior directly. On the other hand, since CRFs relax the assumption of conditional independence of the observed data, they can incorporate global information in the model.

Training data: MRFs can augment small number of expensive labeled data with large number of unlabeled data while CRFs need much labeled data for training.

Data modeling: In MRFs, appropriate distributions need to be selected to model the observed data. In CRFs, good classifier algorithms should be designed for learning from labeled data.

Model selection: At last, our question is which model should be selected in applications. For the tasks of segmentation for video without prior knowledge, like object cutout in video editing [63] and foreground segmentation in surveillance [62], since there is no labeled data or a few interactively labeled data, the MRF model would be selected in common. For the tasks of class labeling problem with large quantities of labeled data, like scene detection in dynamic image sequences [65], the CRF model is used commonly. Actually, the MRF and CRF formulations used in the applications of video segmentation do not strictly comply with the definition of MRFs and CRFs. For instance, the pairwise potential in MRFs is the function of not only the labels but also the intensity difference. In CRFs, the color features are often incorporated in the model by adding the same likelihood term as in MRFs (for example in [66]). This enforces the data independence assumption in CRFs.

3.6 Summary

In summary, this chapter overviews different technologies developed for each step of the pipeline for semantic object segmentation and discusses major challenges at different steps. In order to achieve good performance on semantic object segmentation, local appearance, local consistency, and long-range contextual information need to be considered together. To capture local appearance, filter-banks, visual descriptors and their quantization schemes need to be well designed. They need to have both high discriminative power and good invariance to noise, clutters, and changes of illuminations and viewpoints. Because of the large number of image patches to be processed during object segmentation, computational efficiency is also an important issue to be considered. Conditional random fields provide a powerful framework to integrate local appearance, local consistency and long-range contextual information. However, it requires training data to be labeled at the pixel-level, which is expensive for a large number of object classes. Topic models can learn the models of object classes without supervision or with weak supervision. By including spatial structures, topic models are able to capture long-range contextual information as well as local consistency. However, its capability of modeling local appearance is relatively weak compared with discriminative approaches which use strong classifiers such as SVM and Boosting to model local appearance. It is expected to achieve better performance if the strengths of both generative models and discriminative models can be well combined. For video segmentation, we compare two main statistical frameworks, Markov random fields (MRFs) and conditional random fields (CRFs). The generative approach, MRFs, models the observation data by selecting some conditionally independent distributions. CRFs have better predictive performance since in CRFs the assumption of conditional independency for the observation data is relaxed. But to achieve good enough results, a large number of labeled data should be provided in CRFs. Actually in many real applications, the MRF and CRF model is combined to obtain better results.

References

1. M. Everingham, L. Van Gool, C. K. I. Williams, J. Winn, and A. Zisserman. The PASCAL Visual Object Classes Challenge 2009 (VOC2009) Results. http://www.pascal-network.org/challenges/VOC/voc2009/workshop/index.html.

2. J. Winn, A. Criminisi, and T. Minka. Object categorization by learned universal visual dictionary. In *IEEE International Conference on Computer Vision and Pattern Recognition*, 2005.

3. G. Csurka and F. Perronnin. An efficient approach to semantic segmentation. *International Journal of Computer Vision*, 2010.

4. J. Lafferty, A. McCallum, and F. Pereira. Conditional random fields: Probabilistic models for segmenting and labeling sequence data. In *International Conference on Machine Learning*, 2001.

5. M. Everingham, L. Van Gool, C. K. I. Williams, J. Winn, and A. Zisserman. The PASCAL Visual Object Classes Challenge 2007 (VOC2007) Results. http://www.pascal-network.org/challenges/VOC/voc2007/workshop/index.html.

6. M. Everingham, L. Van Gool, C. K. I. Williams, J. Winn, and A. Zisserman. The PASCAL Visual Object Classes Challenge 2008 (VOC2008) Results. http://www.pascal-network.org/challenges/VOC/voc2008/workshop/index.html.

7. B. C. Russell and A. Torralba. Labelme: a database and web-based tool for image annotation. *International Journal of Computer Vision*, 77:157–173, 2008.

8. Z. Y. Yao, X. Yang, and S. C. Zhu. Introduction to a large scale general purpose groundtruth dataset: Methodology, annotation tool, and benchmarks. In *Proc. Int'l Conf. on EMMCVPR*, 2007.

9. T. Hofmann. Probabilistic latent semantic analysis. In *Proc. of Uncertainty in Artificial Intelligence*, 1999.

10. D. M. Blei, A. Y. Ng, and M. I. Jordan. Latent dirichlet allocation. *Journal of Machine Learning Research*, 3:993–1022, 2003.

11. X. Wang and E. Grimson. Spatial latent dirichlet allocation. In *Proc. Neural Information Processing Systems Conf.*, 2007.

12. T. Leung and J. Malik. Representing and recognizing the visual appearance of materials using three-dimensional textons. *International Journal of Computer Vision*, 43:29–44, 2001.

13. C. Schmid. Constructing models for content-based image retrieval. In *IEEE International Conference on Computer Vision and Pattern Recognition*, 2001.

14. J. Malik, S. Belongie, T. Leung, and J. Shi. Contour and texture analysis for image segmentation. *International Journal of Computer Vision*, 43:7–27, 2001.

15. M. Varma and A. Zisserman. A statistical approach to texture classification from single images. *International Journal of Computer Vision*, 62:61–81, 2005.

16. D. Lowe. Distinctive image features from scale-invariant key points. *International Journal of Computer Vision*, 60:91–110, 2004.

17. K. Mikolajczyk and C. Schmid. A performance evaluation of local descriptors. *IEEE Trans. on Pattern Analysis and Machine Intelligence*, 27:1615–1630, 2005.

18. N. Dalal and B. Triggs. Histogram of oriented gradients for human detection. In *IEEE International Conference on Computer Vision and Pattern Recognition*, 2005.

19. Q. Zhu, M. Yeh, K. Cheng, and S. Avidan. Fast human detection using a cascade of histograms of oriented gradients. In *IEEE International Conference on Computer Vision and Pattern Recognition*, 2006.

20. J. Matas, O. Chum, M. Urban, and T. Pajdla. Robust wide baseline stereo from maximally stable extremal regions. In *Proc. British Machine Vision Conference*, 2002.

21. P. Forssen. Maximally stable colour regions for recognition and matching. In *IEEE International Conference on Computer Vision and Pattern Recognition*, 2007.

22. H. Bay, A. Ess, T. Tuytelaars, and L. van Gool. Surf: Speeded up robust features. *Computer Vision and Image Understanding*, 110:346–359, 2008.

23. S. Lazebnik, S. Schmid, and J. Ponce. A sparse texture representation using local affine regions. *IEEE Trans. on Pattern Analysis and Machine Intelligence*, 27:1265–1278, 2005.

24. K. E. A. Sande, T. Gevers, and G.M. Snoek. Evaluation of color descriptors for object and scene recognition. In *IEEE International Conference on Computer Vision and Pattern Recognition*, 2008.
25. M.A. Tahir, K. Sande, J. Uijlings, F. Yan, X. Li, K. Mikolajczyk, J. Kittler, T. Gevers, and A. Smeulders. Surreyuva-srkda method. In *Pascal VOC 2008 Workshop, Marseille, France*, 2008.
26. E. Nowak, F. Jurie, and B. Triggs. Sampling strategies for bag-of-features image classification. In *Proc. European Conf. Computer Vision*, 2006.
27. F. Jurie and B. Triggs. Creating efficient codebooks for visual recognition. In *Proc. Int'l Conf. Computer Vision*, 2005.
28. D. Nister and H. Stewenius. Scalable recognition with a vocabulary tree. In *IEEE International Conference on Computer Vision and Pattern Recognition*, 2006.
29. F. Moosmann, B. Tigggs, and F. Jurie. Fast discriminative visual codebooks using randomized clustering forests. In *Proc. Neural Information Processing Systems Conf.*, 2006.
30. C. Elkan. Using the triangle inequality to accelerate k-means. In *International Conference on Machine Learning*, 2003.
31. J.C. Van Gemert, J. Geusebroek, C.J. Veenman, and A.W.M. Smeulders. Kernel codebooks for scene categorization. In *Proc. European Conf. Computer Vision*, 2008.
32. J. Shotton, M. Johnson, and Cipolla. Semantic texton forests for image categorization and segmentation. In *IEEE International Conference on Computer Vision and Pattern Recognition*, 2007.
33. P. Geurts, D. Ernst, and L. Wehenkel. Extremely randomized trees. *Machine Learning*, 36: 3–42, 2006.
34. F. Perronnin, C. Dance, G. Csurka, and M. Bressan. Adapted vocabularies for generic visual categorization. In *Proc. European Conf. Computer Vision*, 2006.
35. M. Marszalek and C. Schmid. Accurate object localization with shape masks. In *IEEE International Conference on Computer Vision and Pattern Recognition*, 2007.
36. S. Gould, J. Rodgers, D. Cohen, G. Elidan, and D. Koller. Multi-class segmentation with relative location prior. *International Journal of Computer Vision*, 80:300–316, 2008.
37. D. Cai, X. He, and J. Han. Efficient kernel discriminant analysis via spectral regression. In *Proc. IEEE Int'l Conf. Data Mining*, 2007.
38. D. Aldavert, A. Ramisa, R.L. Mantaras, and R. Toledo. Fast and robust object segmentation with the integral linear classifier. In *IEEE International Conference on Computer Vision and Pattern Recognition*, 2010.
39. X. He, R. S. Zemel, and M. A. Carreira-Perpinan. Multiscale conditional random fields for image labeling. In *IEEE International Conference on Computer Vision and Pattern Recognition*, 2004.
40. J. Shotton, J. Winn, C. Rother, and A. Criminisi. Textonboost for image understanding: Multi-class object recognition and segmentation by jointly modeling texture, layout, and context. *International Journal of Computer Vision*, 81:2–23, 2009.
41. A. Torralba, K. P. Murphy, and W. T. Freeman. Sharing visual features for multiclass and multi-view object detection. *IEEE Trans. on Pattern Analysis and Machine Intelligence*, 19:854–869, 2007.
42. B.A. Fulkerson, A. Vedaldi, and S. Soatto. Class segmentation and object localization with superpixel neighborhoods. In *Proc. Int'l Conf. Computer Vision*, 2009.
43. X. Ren and J. Malik. Learning a classification model for segmentation. In *Proc. Int'l Conf. Computer Vision*, 2003.
44. X. He, R. S. Zemel, and D. Ray. Learning and incorporating top-down cues in image segmentation. In *Proc. European Conf. Computer Vision*, 2006.
45. A. Torralba, K.P. Murphy, and W. Freeman. Contextual models for object detection using boosted random fields. In *Proc. Neural Information Processing Systems Conf.*, 2004.
46. A. Rabinovich, A. Vedaldi, C. Galleguillos, E. Wiewiora, and S. Belongie. Objects in context. In *Proc. Int'l Conf. Computer Vision*, 2007.
47. A. Quattoni, M. Collins, and T. Darrell. Conditional random fields for object recognition. In *Proc. Neural Information Processing Systems Conf.*, 2004.

48. X. Ma and W.E.L. Grimson. Learning coupled conditional random field for image decomposition with application on object categorization. In *IEEE International Conference on Computer Vision and Pattern Recognition*, 2008.
49. J. Reynolds and K. Murphy. Figure-ground segmentation using a hierarchical conditional random field. In *Proc. of Canadian Conference on Computer and Robot Vision*, 2007.
50. J. Sivic, B. C. Russell, A. A. Efros, A. Zisserman, and W. T. Freeman. Discovering object categories in image collections. In *Proc. Int'l Conf. Computer Vision*, 2005.
51. B. C. Russell, A. A. Efros, J. Sivic, W. T. Freeman, and A. Zisserman. Using multiple segmentations to discover objects and their extent in image collections. In *IEEE International Conference on Computer Vision and Pattern Recognition*, 2006.
52. J. Verbeek and B. Triggs. Region classification with markov field aspect models. In *IEEE International Conference on Computer Vision and Pattern Recognition*, 2007.
53. J. Verbeek and B. Triggs. Scene segmentation with conditional random fields learned from partially labeled images. In *Proc. Neural Information Processing Systems Conf.*, 2007.
54. T. L. Griffiths and M. Steyvers. Finding scientific topics. In *Proc. of the National Academy of Sciences of the United States of America*, 2004.
55. L. Fei-Fei, R. Fergus, and P. Perona. Learning generative visual models from few training examples: an incremental bayesian approach teseted on 101 object categories. In *in Proc. IEEE CVPR Worshop of Generative Model Based Vision*, 2004.
56. T. S. Ferguson. A bayesian analysis of some nonparametric problems. *The Annals of Statistics*, 1:209–230m, 1973.
57. J Shi and J. Malik. Normalized cuts and image segmentation. *IEEE Trans. on Pattern Analysis and Machine Intelligence*, 22:888–905, 2000.
58. D. Larlus, J. Verbeek, and F. Jurie. Category level object segmentation by combining bag-of-words models with dirichlet processes and random fields. *International Journal of Computer Vision*, 88:238–253, 2010.
59. G. Passino, I. Patras, and E. Izquierdo. Latent semantics local distribution for crf-based image semantic segmentation. In *Proc. British Machine Vision Conference*, 2009.
60. E. B. Sudderth, A. Torralba, W. T. Freeman, and A. S. Willsky. Describing visual scenes using transformed objects and parts. *International Journal of Computer Vision*, 77:291–330, 2007.
61. L. Cao and L. Fei-Fei. Spatially coherent latent topic model for concurrent object segmentation and classification. In *Proc. Int'l Conf. Computer Vision*, 2007.
62. J. Sun, W. Zhang, X. Tang, and H. Shum. Background cut. In *Proc. European Conf. Computer Vision*, 2006.
63. Y. Boykov and M. Jolly. Interactive graph cuts for optimal boundary and region segmentation of objects in nd images. In *Proc. Int'l Conf. Computer Vision*, 2002.
64. A. Criminisi, G. Cross, A. Blake, and V. Kolmogorov. Bilayer segmentation of live video. In *IEEE International Conference on Computer Vision and Pattern Recognition*, 2006.
65. C. Wojek and B. Schiele. A dynamic conditional random field model for joint labeling of object and scene classes. In *Proc. European Conf. Computer Vision*, 2008.
66. P. Yin, A. Criminisi, J. Winn, and M. Essa. Tree-based classifiers for bilayer video segmentation. In *IEEE International Conference on Computer Vision and Pattern Recognition*, 2007.
67. Y. Wang and Q. Ji. A dynamic conditional random field model for object segmentation in image sequences. In *IEEE International Conference on Computer Vision and Pattern Recognition*, 2005.

Chapter 4
Video Scene Analysis: A Machine Learning Perspective

Wen Gao, Yonghong Tian, Lingyu Duan, Jia Li, and Yuanning Li

Abstract With the increasing proliferation of digital video contents, learning-based video scene analysis has proven to be an effective methodology for improving the access and retrieval of large video collections. This chapter is devoted to present a survey and tutorial on the research in this topic. We identify two major categories of the state-of-the-art tasks based on their application setup and learning targets: generic methods and genre-specific analysis techniques. For generic video scene analysis problems, we discuss two kinds of learning models that aim at narrowing down the *semantic gap* and the *intention gap*, two main research challenges in video content analysis and retrieval. For genre-specific analysis problems, we take sports video analysis and surveillance event detection as illustrating examples.

4.1 Introduction

With the increasing proliferation of digital video contents, efficient techniques for analysis, indexing and retrieval of videos according to their contents become more and more important. In general, the analysis of video sequences involves a wide spectrum of techniques from low-level content analysis such as feature extraction, structure analysis, object detection and tracking, to high-level semantic analysis such as scene analysis, event detection, and video mining. There has been a lot of progress made in each of the modules in the above pipeline. We refer the readers to existing texts and reviews for video content analysis [11,26,60] for a comprehensive treatment.

Scene analysis and understanding plays an important role in video content analysis and semantic-based video retrieval, since a scene usually contains a collection of semantically related and temporally adjacent shots, depicting and conveying a high-level concept or story that users are mostly interested in. In general, scene understanding may involve, understanding the scene structure (e.g. pedestrian

W. Gao (✉)
School of EE & CS, Peking University, Beijing 100871, China
e-mail: wgao@pku.edu.cn

K.N. Ngan and H. Li (eds.), *Video Segmentation and Its Applications*,
DOI 10.1007/978-1-4419-9482-0_4, © Springer Science+Business Media, LLC 2011

sidewalks, east-west roads), scene status (e.g. traffic jam, crowd), scene categories or concepts (e.g. street, forest), scene motion patterns (e.g. vehicles making u-turns, north–south traffic), etc. [66]. With the knowledge of scene structure, categories, activities and motion patterns, low-level tracking and abnormal activity detection can be improved and high-level event analysis and video retrieval can be accomplished.

The usage of machine learning techniques has proven to be a robust methodology for semantic scene analysis and understanding. The main characteristic of learning-based approaches is their ability to adjust their internal structure according to input and respective desired output data pairs in order to approximate the relations implicit in the provided (training) data, thus elegantly simulating a reasoning process. Consequently, learning methods constitute an appropriate solution for scene analysis when the considered a-priori knowledge cannot be defined explicitly [53]. Various learning techniques, such as Bayesian networks (BNs) [1], latent aspect models [3,40], linear discriminant analysis [45] and support vector machines (SVMs) [69], are widely adopted to map data representations to semantic descriptions.

We devote this chapter to present a survey and tutorial on the problems and solutions of video scene analysis, put in the perspectives of the learning components and tasks. We identify two major categories of the state-of-the-art tasks based on their application setup and learning targets: generic methods and genre-specific analysis techniques. For generic video scene analysis problems, we discuss two kinds of learning models that aim at narrowing down both *semantic gap* and *intention gap*, two main research challenges in video content analysis and retrieval. For genre-specific analysis problems, we take sports video scene analysis and surveillance event detection as illustrating examples. Clearly, our review of specific approaches is by no means complete, in part due to the rapid development in the area.

Modeling scene semantics and learning them from video data are of interest to numerous research areas, including multimedia retrieval, machine learning, pattern recognition, computer vision, knowledge representation, and data mining. Problems in this area provide synergies among these areas for the understanding of video content. The underlying data processing and learning methodology used here are very similar to those seen in many other domains such as text analysis and data mining, but present new challenges such as cross-media correlation and temporal indexing. Moreover, with the ubiquitous presence of video data in our lives, better modeling and learning semantics of video scenes will enable better user experiences and improve system design in closely related areas such as multimedia resource management and retrieval.

4.2 Description of Scene Semantics

One of the main difficulties inherent in the video scene analysis is the richness of semantic content interpretable within a scene. We take Fig. 4.1 as an example to illustrate the complexity of scene semantics. A human observer can easily deduce

Semantic Facets		Learning Task
Which	**Category:** outdoor/city	Scene categorization
What	**Generic Object/Scene:** crowd, flag, buildings **Specific Object/Scene:** B. Obama, M. Obama, dais	Automatic annotation, Object detection & recognition
Where	**Spatial Relationship:** M. Obama *<Left-of>* B. Obama	Object localization
How	**Activity:** waving **Event:** assembly	Behavior recognition, Event detection

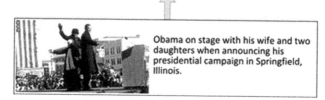

Obama on stage with his wife and two daughters when announcing his presidential campaign in Springfield, Illinois.

Fig. 4.1   An example of the faceted representation of scene semantics

many semantic information only from this scene: four persons (a man, a woman and two children) are waving in an assembly. With contextual knowledge, one can further know this scene is about B. Obama with his wife and two daughters when announcing his presidential campaign in Springfield, Illinois, USA. However, the semantic descriptions that can be automatically inferred by a learning system are very limited. For example, the scene might be categorized into "outdoor" or "city" by image classification algorithms or annotated as "*crowd, flag, building*" by automatic annotation models; the person in the scene might be recognized as "B. Obama" by face recognition algorithms; we can also use object localization algorithms to learn the spatial relationship of objects (e.g., B Obama *Center-of* the picture); furthermore, high-level concept detection algorithms can be used to detect activities (e.g., "*waving*") or events (e.g., "*assembly*"). As shown in Fig. 4.1, these semantics can be summarized along the four aspects–*which* (semantic types or categories), *what* (objects or scenes), *where* (spatial relationships), and *how* (actions, activities or events):

1. *Which – Semantic Types and Categories*: The *which* facet typically refers to semantic types or categories of scenes. Given a taxonomy, this facet helps answer the question: *which type or category is the scene?* Description of scene semantics using the *which* facet is very general, but prove to be of great importance for either organizing unseen images/scenes into broad categories, or for semantic-based retrieval from large-scale collections.

2. *What – Objects and Scenes*: The *what* facet describes the objects and scenes in an image/video. It answers the question *what is the subject (object/scene, etc.) in it?* Extracting the *what* facets from scenes covers a wide range of visual learning

tasks. Typically, the generic *what* description of a scene can be extracted by automatic annotation, while the specific description can be produced by object detection and recognition such as face recognition.

3. *Where – Spatial Relationships*: The *where* facet often refers to the locations of objects or spatial relationships between objects in an image/video. It is the answer to the questions *where is the object in a scene? or where the objects are placed relative to each other?* The problem of extracting this *where* facet is challenging because objects of the same type may appear in different locations, scales, or under occlusions and deformations. Nevertheless, the localization of objects can then be used to address a range of tasks, including descriptive classification, search, and clustering.

4. *How – Actions, Activities, and Events*: The *how* facet characterizes high-level abstract concepts that are expressed in images/videos. It answers the question *how about the subject in the scene? or what is happening in this scene?* In general, the *how* facet can be extracted by different high-level concept detection algorithms, depending on the data available and the target decision. We often need to elaborately design a detection algorithm with the specific feature representation for each of the actions, activities, and events.

Characterization of scene semantics is the basis for setting up learning problems and defines goals of such systems. Such a faceted representation of scene semantics can be used to clarify assumptions and targets in learning tasks, because these facets are key attributes that are sufficient and necessary to summarize the semantic content of a scene, and also because it is possible to extract them from images or videos by using machine learning techniques.

4.3 Generic Techniques for Video Scene Analysis

This section will discuss the main challenges for research on video scene analysis, and then present two representative works on learning-based video scene analysis, one for bridging the semantic gap with automatic annotation while the other for capturing user's visual attention patterns with visual saliency learning.

4.3.1 Research Challenges

The learning problems for video scene analysis exhibit a large variability. Generally speaking, the problems are essentially influenced by the properties of the target semantics, constrained by the availability of data, and directed by the goals of the intended tasks and applications. Figure 4.2 summarizes two main research challenges in video content analysis and retrieval, which should been addressed in the learning problems for video scene analysis.

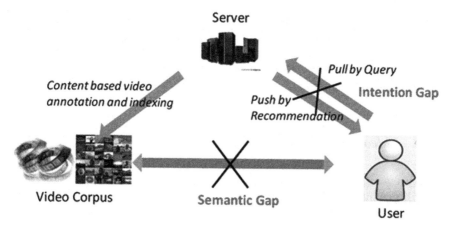

Fig. 4.2 An illustration of two research challenges in video content analysis and retrieval

Generally speaking, the main challenge of video content analysis is understanding media by bridging the *semantic gap* between the bit stream on the one hand and the visual content interpretation by humans on the other. Clearly, the semantic gap is a fundamental problem in multimedia analysis and indexing that almost all research papers in the field must address [50]. As a computational problem, the semantic gap is tightly related to the modeling and analysis of video scenes. As pointed out by Hare et al. [14], the semantic gap can be further divided into two major sections: the gap between the low-level descriptors and object labels, and the gap between the labeled objects and the full semantics. For reasons that will become clear later, we refer to them respectively as the *which–what* gap and the *where–how* gap. It is should be noted that although expressed by the two designations, the full connotations of the two sections of the semantic gap are much broader. In Sect. 4.3.2, we will present a general framework for capturing various facets of scene semantics by taking into account both temporal and spatial contexts.

Another research challenge that has received much attention of research in the multimedia retrieval field is a gap between users' search intents and the queries, called *intention gap* [67]. Due to the incapability of keyword queries to express users' intents, intention gap often leads to unsatisfying search results. Despite originated from multimedia retrieval, this gap may exist in a boarder range of multimedia applications such as user-targeted video advertising and content-based filtering. For example, a less intrusive model of advertising is only displaying advertising information when the user makes the choice by clicking on an object in a video. Since it is the user who requests the product information, this type of advertising is better targeted and likely to be more effective. By learning user's visual attention patterns, the hot-spots that correspond to brands can be further highlighted so as to extract more attention of users.

Often, visual attention is operationalized as a selection mechanism to filter out unwanted information in a scene [21]. By focusing on the attractive region, a scene can be analyzed in a user-targeted manner. Generally, determining which region will capture attention requires finding the value, location, and extent of the most salient subset of the input visual stimuli. Typically, saliency reveals the probability that a location is salient and involves the stimulus-driven (i.e., bottom-up) and task-related (i.e., top-down) components in human vision system. In Sect. 4.3.3, we will present two rank learning approaches for visual saliency estimation in video.

4.3.2 Video Annotation with Sequence Multilabeling

In recent years, various supervised learning methods (e.g. support vector machines [8, 65], graphical models [41] and multi-modality fusion methods [27, 51]) are employed to find out the informative feature patterns to detect concepts within video data. However, due to the well-known semantic gap, video annotation methods purely relying on low-level features only couldn't achieve the desirable performance.

Video data are by nature rich in spatial and temporal context that could be useful to facilitate annotation. Generally speaking, semantic concepts may have *spatial correlations* within a shot and *temporal consistencies* between consecutive shots. That is, several concepts may co-occur within a shot due to the spatial correlation, and a concept could be persistent across several neighboring shots due to the temporal consistency. Taking an example in Fig. 4.3, *street* and *building* co-occur in shot$_t$ and shot$_{t+1}$, while car is present in three consecutive shots. Moreover, it is noted that two distinct concepts may correlate with each other between shots. This

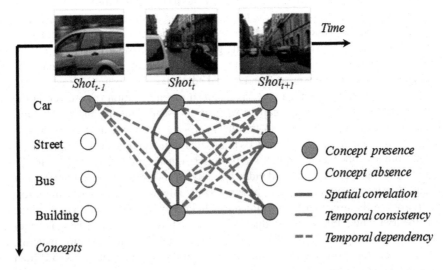

Fig. 4.3 Illustration of video annotation with multilabels. For a shot sequence, concepts present in neighboring shots exhibit several contextual relationships. Note that here both temporal consistency of a concept and temporal dependency between concepts across neighboring shots are referred to as temporal correlation

contextual relationship can be denoted as *temporal dependency*. In Fig. 4.3, when car is present in $shot_{t-1}$, *street* would probably be detected in next two shots. In this section, such interaction of concepts within the same shot is referred to *spatial correlation*; the temporal consistency of a concept and the temporal dependency between distinct concepts are jointly referred to as *temporal correlation*. Both spatial and temporal correlations of concepts will be exploited to facilitate video annotation in our proposed sequence multilabeling framework.

In this section, video annotation is formulated as a sequence multilabeling (SML for short) problem, where a sequence of multilabels is predicted simultaneously for a set of consecutive video shots given a list of predefined concepts. This is different from most existing video annotation paradigms working on individual shots. SML provides a unified video annotation framework to incorporate spatial and temporal context. Accordingly, learning algorithms seek to capture informative features and contextual correlations of concepts so as to facilitate video annotation. To address the SML problem, a novel discriminative method, called *sequence multilabel support vector machine* (SVM^{SML} for short), is proposed. In SVM^{SML}, a joint kernel is employed to model three relationships (i.e., dependencies of labels on low-level overlapping features, spatial and temporal correlations of labels). Experiments on TRECVID'05 and TRECVID'07 datasets have shown that SVM^{SML} gains superior performance over state-of-the-art methods.

4.3.2.1 Previous Work

Video annotation is traditionally formulated as a multilabeling problem over individual shots, which is referred to *individual multilabeling* (IML) in this section. IML treats video shots as independent instances, where either multiple concept detectors [52, 65] or a multilabel classifier [48] are learned at the shot level. By exploiting spatial correlations of concepts within individual shots (as shown in Fig. 4.4a), many research efforts have been devoted to enhance the performance of IML. Smith et al. [51] propose a two-stage discriminative fusion method to explore the concept correlation within a shot. Alternatively, graphical methods (e.g., Bayesian network [41, 42], conditional random fields [26, 63] and graph diffusion [27]) have been employed to model the spatial correlations and to refine the annotation results. In addition, re-ranking approaches [17, 28, 43] have attracted much attention. For example, Kennedy et al. [28] use random walk to exploit the contextual correlation

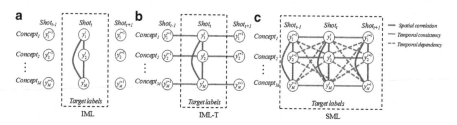

Fig. 4.4 Three paradigms of video annotation

of features. Recently, Qi et al. [48] employ a multilabel classifier to build up a correlative multilabeling (CML) framework. CML employs the concept correlations within individual shots and annotates shots with multiple concepts simultaneously. Overall, these works detect concepts within individual shots independently. In a sense, they can be considered as a sort of direct extension of image annotation methods to video domain, with much less temporal context taken into account.

As video is temporally informative, researches attempt to utilize temporal information to enhance video annotation. Generally speaking, video annotation methods over sequential shots can be categorized into three types. The first type is to model the temporal pattern of low-level features. For example, hidden Markov model (HMM) is employed by Xie et al. [59] to model the temporal dynamics of low-level features (e.g. color and motion) for specific video event detections. Qi et al. [48] introduce a temporal kernel into CML to model the similarity between sequences of low-level feature. In these works, the temporal dynamic in low-level feature is employed to improve specific concept detectors whereas higher-level temporal correlations of concepts are ignored.

The second type is to perform temporal refinement over IML (IML-T), in which concepts are first annotated over individual shots followed by the refinement with temporal consistency (as shown in Fig. 4.4b). For example, Yang et al. [64] and Liu et al. [35] incorporate temporal consistency into active learning to detect multiple video concepts. Weng et al. [58] and Liu et al. [36] propose several fusion methods to refine the annotation results of individual shots, where spatial correlations and temporal consistencies of concepts are modeled by association rules and temporal filtering respectively. Higher-order temporal consistency of concept is also explored in [58]. In these methods, the outputs of each concept detector across consecutive shots are smoothed to keep temporal consistency. However, the pairwise interactions of distinct concepts across consecutive shots are ignored basically. Despite more or less improvements, these methods are weak for unstable performance.

The third type is to model the spatial and temporal context of concepts temporally. Besides the dynamics of low-level features, spatial and temporal contexts of higher-level concepts would be useful to assist event/action detection [9, 57]. However, there are few generic approaches in enhancing video annotation with spatial and temporal correlations of concepts. Naphade et al. [41] try to integrate spatial co-occurrence and temporal dependency of concepts into a probabilistic Bayesian network so that the pair-wise relationships of concepts from one frame (or shot) and between two adjacent frames (or shots) can be modeled. Alternatively, in this section, video annotation is formulated as a sort of sequence multilabeling and solved with a unified learning framework to capture both spatial and temporal correlations of semantic concepts (as shown in Fig. 4.4c). Compared with IML-T methods, SVM^{SML} learns both SML score function and the contributions of multiple cues (i.e., distinct low-level features, spatial and temporal correlations of concept labels) in one single stage over the same training dataset. SVM^{SML} do not require any initial annotation and has greatly alleviated the problem of error propagation. Also learning the SML score function as well as spatial and temporal context over the same training data avoids additional efforts on data collection and labeling. Compared with [41], SVM^{SML} is a kernel-based method where the spatial correlation, the first and the second order temporal correlations are modeled within a joint kernel. Moreover, multimodal features and temporal dynamics of the low-level feature can be integrated into SVM^{SML} in the manner of basic kernels.

4.3.2.2   SML Formulation for Video Annotation

Let $\mathbf{x} = (\mathbf{x}^1,\ldots,\mathbf{x}^t,\ldots,\mathbf{x}^T) \in \chi$ denote the sequence of input features (i.e. visual/audio/text features) extracted from a video clip consisting of T shots, where χ is the input feature space. The output sequence of multilabels is expressed by $\mathbf{y} = (\mathbf{y}^1,\ldots,\mathbf{y}^t,\ldots,\mathbf{y}^T) \in \kappa$, where $\mathbf{y}^T \in v$. Here v and κ are the output spaces of individual shot and shot sequence, respectively. Let $C = (c_1,\ldots,c_m,\ldots,c_M)$ represent the lexicon of M semantic concepts. Each entry (i.e. the multilabel of an elementary shot) of the output multilabel sequence can be expressed by an M dimensional label vector $\mathbf{y}^t = (y_1^t,\ldots,y_m^t,\ldots,y_M^t)'$, where $y_m^t \in \{1,0\}$ indicates whether concept c_m is present in the tth shot. Accordingly, $L = \{(\mathbf{x}_1,\mathbf{y}_1),\ldots,(\mathbf{x}_i,\mathbf{y}_i),\ldots,(\mathbf{x}_N,\mathbf{y}_N)\}$ denotes the training set consisting of N sequences.

Given the training set L, SML aims to learn an optimal mapping from a sequence of input features to a sequence of output multilabels. For an unknown shot sequence $\mathbf{x}$, the sequence of output multilabels can be predicted as:

$$\begin{aligned} \mathbf{y}^* &= (\mathbf{y}^{1*},\ldots,\mathbf{y}^{t*},\ldots,\mathbf{y}^{T*}) \\ &= \arg\max_{(y^1,\ldots,y^T)\in\kappa} F(\mathbf{x}^1,\ldots,\mathbf{x}^t,\ldots,\mathbf{x}^T,\mathbf{y}^1,\ldots,\mathbf{y}^t,\ldots,\mathbf{y}^T), \end{aligned} \quad (4.1)$$

where $F(\cdot)$ is SML score function over the input feature sequence and the output multilabel sequence. SML predicts the annotation of the shot sequence by maximizing the score function $F(\cdot)$ over all candidate multilabel sequences. As shown in Fig. 4.4c, different types of spatial and temporal contexts in the shot sequence can be also incorporated with the prediction.

SML is a generalized formulation for video annotation. That is, IML and IML-T can be viewed as two special cases of SML. When all video shots are assumed to be independent with each other, SML reduces to IML:

$$\mathbf{y}^* = (\mathbf{y}^{1*},\ldots,\mathbf{y}^{t*},\ldots,\mathbf{y}^{T*}), \text{ where } \mathbf{y}^{t*} = \arg\max_{y^t \in v} F(\mathbf{x}^t,\mathbf{y}^t). \quad (4.2)$$

In IML, detection of one concept only depends on low-level features and other concepts within current shot.

IML-T is a two-step optimization process which improves the initial annotation results of IML by:

$$\mathbf{y}^{t*} = \arg\max_{y^t \in v} \varphi\left(\mathbf{y}^{(t-w)'},\ldots,\mathbf{y}^{(t-1)'},\mathbf{y}^{t'},\mathbf{y}^{(t+1)'},\ldots,\mathbf{y}^{(t+w)'}\right)$$

$$\text{where } \mathbf{y}^{t'} = \arg\max_{y^t \in v} F(\mathbf{x}^t,\mathbf{y}^t). \quad (4.3)$$

In (4.3), w corresponds to the neighborhood size of current shot and φ characterizes the temporal consistencies of individual concepts within the neighborhood (as shown in Fig. 4.4b). IML-T improves the initial detection of one concept in current shot by smoothing the annotation results within the neighborhood.

We can further explain their differences from the optimization perspective. Given a sequence of correlated shots (e.g. a scene), IML-based approaches can provide local optimal annotation results within each shot; IML-T based approaches improve the IML results by temporal consistency within neighboring shots; while SML-based approaches can effectively find out the near-optimal annotation in the sense of shot sequence.

4.3.2.3 SVM^{SML}: A Discriminative Method for SML

SML is to predict output that is not simple binary label, but instead has a more complex sequential multilabel structure. By appropriately modeling multiple relationships (i.e., the dependencies of labels on overlapping low-level features, spatial and temporal correlation of labels), we extend a machine learning method called structure SVM (SVM^{struct}) [55] and propose SVM^{SML} to learn a structured score function which can make better use of the available training data. Similar to other structure SVM$_s$, SVM^{SML} learns a discriminative score function $F : \chi \times \kappa \to \mathbb{R}$ over input/output pairs. In SVM^{SML} framework, the multilabel sequence is then predicted by (4.1), which maximizes the response score of F over the output space κ for a given input $\mathbf{x}$. The proposed SML score function is linear in combining the joint feature representation $\Phi(\mathbf{x}, \mathbf{y})$:

$$F(\mathbf{x}, \mathbf{y}; \mathbf{w}) = \langle \mathbf{w}, \Phi(\mathbf{x}, \mathbf{y}) \rangle, \tag{4.4}$$

$\mathbf{w}$ is a vector of linear combination weights, which will be optimized by solving the dual problem of SVM^{SML}; $\Phi(\mathbf{x}, \mathbf{y})$ is a joint feature representation w.r.t. input and output pair. As a variant of SVM, SVM^{SML} employs kernel function to compute the inner product in the joint feature space. Accordingly, we can explicitly define a similarity measure between two shot sequences and implicitly map original feature space to a high dimensional feature space, thereby avoiding explicit feature representation of $\Phi(\mathbf{x}, \mathbf{y})$ and the curse of dimension [2]. SVM^{SML} holds some basic advantages, in particular the generalization abilities that have been theoretically guaranteed by margin-maximization property of the learning algorithm [4].

Figure 4.5 illustrates the framework of our approach. At the training phase, low-level multimodal features (e.g. visual features such as color, texture and shape, text features from ASR/OCR transcripts, or audio features) and the multilabel sequences from training shot sequences are fed into a joint kernel. This joint kernel models the dependencies of labels on overlapping low-level features, spatial correlations of labels within a shot, as well as temporal correlations of labels from consecutive shots in a linear combination of kernels. SVM^{SML} is learnt by a working set optimization method, where kernel weights of the joint kernel are simultaneously

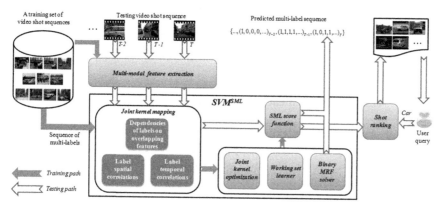

Fig. 4.5 The sequence multilabeling framework for video annotation

Table 4.1 Experimental results of SVM^{SML} and several state-of-the-art methods

Dataset	Annotation method	Concept number	MAP
TRECVID'05	CML [48]	39	0.2901
	CML-T [48]		0.3325
	GD [27]		0.1858
	SVM^{SML}		**0.5065**
	CRF [25]	26	0.5001
	SVM^{SML}		**0.5573**
TRECVID'07	OWA [31]	20	0.1320
	MLMK [65]		0.3325
	SVM^{SML}		**0.3147**

optimized by a MKL-based learning algorithm. At the testing phase, one multilabel sequence is predicted by the learnt SML score function for the testing shot sequence. During the training and testing phases, a BMRF-based approximate method is employed to accelerate the search process over the large output space of multilabel sequence. Finally, testing shots can be ranked by their relevance scores to a given query concept. It is worthy to note that such framework can work at the finer granularities (e.g., sub-shots or sampled frames with equal interval). Without loss of generality, shot sequences are studied here.

In Table 4.1, performance of SVM^{SML} is evaluated with Mean Average Precision (MAP) over TRECVID'05 and TRECVID'07 datasets. It is shown that SVM^{SML} gains superior performance over several state-of-the-art methods [25, 26, 31, 48, 65].

4.3.3 Visual Saliency Estimation Using Rank Learning

In this section we demonstrate the usage of machine learning approaches in visual saliency estimation. We begin by introducing the concept and problems in visual saliency estimation, followed by a brief review of related works. After that, we discuss two specific methods that estimate visual saliency by using the rank learning approaches.

4.3.3.1   What Is Visual Saliency Estimation?

In natural scenes, the complexity of the input visual stimuli usually exceeds the processing capacity of human vision system. As a consequence, the important visual subsets will be selected and processed with higher priorities. In this selective mechanism, visual saliency often plays an essential role in determining which subset (e.g., pixel, block, region or object) in a scene is important. Therefore, the central task in visual saliency estimation is to rank various visual subsets in a scene to indicate their importance and processing priorities.

Typically, saliency involves the stimulus-driven (i.e., bottom-up) and task-related (i.e., top-down) components in human vision system. The bottom-up component comprises of the low-level processes and is driven by the intrinsic attributes of the stimuli. In contrast, the top-down component involves the high-level processes in which the deployment of attention can be modulated by task, consequently demonstrating a biased selectivity on the input stimuli [21]. Therefore, visual attention analysis requires modeling visual saliency by simultaneously taking the bottom-up and top-down factors into account.

4.3.3.2   Previous Work

Typically, most of existing bottom-up saliency models select the unique or rare subsets in a scene as the salient regions. Generally speaking, these approaches were guided by the Feature Integration Theory [54], which posited that different features can be bound into consciously experienced wholes for visual saliency estimation. For example, Itti et al. [24] proposed a set of preattentive features including center-surround intensity, color, and direction contrasts. These contrasts were then integrated to compute image saliency through the winner-take-all competition. In [19], this framework was extended to video saliency by introducing motion and flicker contrasts. Using the same features, Itti and Baldi [20] recovered video saliency as "surprise" by combining spatial contrast and temporal evolution. Walther and Koch [56] extended the framework in [24] to salient "proto-objects," which were described as volatile, bottom-up units that could be bound into objects if attended to. Similarly, Hu et al. [18] assigned high saliency to the image block with high texture contrast.

Different from these contrast-based approaches, Bruce and Tsotsos [5] established a bottom-up model based on the principle of maximizing information sampled from a scene. In [68], video saliency was computed according to the motion irregularity derived from inter-frame key points matching. Harel et al. [15] built a graphical model to compute image saliency. A random walker was adopted on the graph and the less-visited nodes (pixels) were selected as "salient". Using spectrum analysis, Hou and Zhang [16] computed image saliency by extracting spectral residuals in the amplitude spectrum of Fourier Transform, and Guo et al. [13] modeled video saliency as a sort of inconsistency in the phase spectrum of Quaternion Fourier Transform. Alternatively, Marat et al. [38] presented a biology-inspired spatiotemporal saliency model. The model extracted two signals from video stream corresponding to the two main outputs of the retina. Both signals were then transformed and fused into a spatiotemporal saliency map.

From the phenomenological perspective, the bottom-up approaches estimate visual saliency mainly based on the visual stimuli. However, the task which involves an act of "will" on the probable salient targets also plays an important role. Biological evidences show that the neurons linked with various stimuli undergo a mutual competition to generate the bottom-up saliency, while the task can bias such competition in favor of a specific category of stimuli [10]. For example, Peters and Itti [46] showed that when performing different tasks in video games, individual's attention could be predicted by respective task-relevant models. In these processes, the adopted tasks worked as different top-down controls to modulate the bottom-up process. In real-world scenes, however, it is difficult to explicitly predefine such tasks. Instead, some approaches such as [18] and [32] treated the top-down control as the priors to segment images before the bottom-up saliency estimation. In their works, an image was first partitioned into regions, and the regional saliency was then estimated by regional difference. Other works, such as [6] and [37], introduced the top-down factors into the classical bottom-up framework by extracting semantic clues (e.g., face, speech and music, camera motion). These approaches could provide impressive results but relied on the performance of image segmentation and semantic clue extraction.

Recently, machine learning approaches have been introduced in modeling visual saliency to learn the top-down control from recorded eye-fixations or labeled salient regions. Typically, the top-down control works as a "stimulus-saliency" function to select, re-weight and integrate the input visual stimuli. For example, Itti and Koch [23] proposed a supervised approach to learn the optimal weights for feature combination, while Peters and Itti [47] presented an approach to learn the projection matrix between global scene characteristics and eye density maps. Navalpakkam and Itti [44] modeled the top-down gain optimization as maximizing the signal-to-noise ratio (SNR). That is, they learned linear weights for feature combination by maximizing the ratio between target saliency and distractor saliency. Besides learning the explicit fusion functions, Kienzle et al. [29] proposed a nonparametric approach to learn a visual saliency model from human eye-fixations on images. A support vector machine (SVM) was trained to determine the saliency using the local intensities. For video, Kienzle et al. [30] presented an approach to learn a set of temporal filters from eye-fixations to find the interesting locations. On the regional saliency dataset, Liu et al. [33] proposed a set of novel features and adopted a conditional random field (CRF) to combine these features for salient object detection. After that, they extended the approach to detect salient object sequences in video [34].

To sum up, it is highly useful to incorporate the learning-based top-down control into the visual saliency model. In the remainder of this section, we will discuss two specific methods that estimate visual saliency by using the rank learning approaches.

4.3.3.3 Cost-Sensitive Rank Learning for Visual Saliency Estimation in Video

In learning-based visual saliency estimation, the users' eye traces are usually obtained to construct the ground-truth training data. That is, the eye traces can reveal whether certain locations are salient enough to attract human visual attention. However, these eye traces can only provide sparse positive samples in video data since each video frame can be displayed with quite a short time. Consequently, only a few locations in a video scene can be labeled by eye fixations as positive, while most of other locations in the scene remain unlabeled (as shown in Fig. 4.6a). These unlabeled data may contain many positive samples so that it is improper to treat all of them as negative samples or randomly select negative samples from them.

To solve this problem, we propose a cost-sensitive rank learning approach on positive and unlabeled data for visual saliency estimation. In our approach, we avoid

Fig. 4.6 Generating ground-truth saliency from sparse positive samples. (**a**) Sparse positive samples (i.e., the eye fixations); (**b**) the visual similarity map; (**c**) the spatial correlation map; (**d**) the derived ground-truth map

the explicit extraction of positive and negative samples by directly integrating both the positive and unlabeled data into the optimization objective in a cost-sensitive manner. In this process, we first recover the ground-truth saliency maps from the limited eye fixations received by each frame. The basic principle is that visual subsets that are adjacent and similar to the eye fixations should be assigned with high saliency values. Toward this end, the visual similarity map is calculated to pop-out the locations that are similar to the positive samples (as shown in Fig. 4.6b), while the spatial correlation map is derived to pop-out the neighbors of the eye fixations (as shown in Fig. 4.6c). Finally, the visual similarity map and the spatial correlation map are combined to derive the ground-truth saliency map (as shown in Fig. 4.6d). For the sake of convenience, the ground-truth saliency values are normalized into $[0, 1]$.

With the ground-truth saliency values, we train a ranking function $\varphi(\mathbf{x}) = \omega^T \mathbf{x}$ that can integrate various local visual attributes (represented by a column vector $\mathbf{x}$) for visual saliency estimation. For two locations $\mathbf{B}_{km}$ and $\mathbf{B}_{kn}$ with ground-truth saliency values g_{km} and g_{kn}, $\varphi(\mathbf{x}_{km}) > \varphi(\mathbf{x}_{kn})$, indicates that $\mathbf{B}_{km}$ ranks higher than $\mathbf{B}_{kn}$ and maintains a higher saliency. In the training process, it is often difficult to directly determine the label for each training sample, especially for the one with medium ground-truth saliency (e.g., around 0.5). Therefore, we integrate all the positive and unlabeled data into a rank learning framework in a cost-sensitive manner. Toward this end, the empirical loss can be defined as:

$$L(\omega) = \sum_k \sum_{m \neq n} [g_{km} - g_{kn}]_+ \left[\omega^T \mathbf{x}_{km} \leq \omega^T \mathbf{x}_{kn} \right]_1 \qquad (4.5)$$

Where $[x]_+ = \max(0, x)$. Note that here $[E]_1 = 1$ if event E holds, otherwise $[E]_1 = 0$. We can see that there will be a loss if the ranking function gives predictions contrary to the ground-truth saliencies. Moreover, the loss emphasizes the correlations between targets and distractors since the central issue in visual saliency estimation is to distinguish targets from distractors. That is, the cost of erroneously ranking a target-distractor pair (i.e., $g_{km} - g_{kn} \to 1$) is much bigger than that of mistakenly predicting the ranks between target pairs or between distractor pairs (i.e., $g_{km} - gkm \to 0$). Thus it is cost-sensitive by differentiating target-distractor pairs in our framework.

Often, it is difficult to minimize such a loss with binary terms. Thus we simply replace each binary term with its upper bound (e.g., exponential upper bound) and obtain a convex optimization objective. After that, the global optimum can be reached using gradient-based method and we can obtain the optimal ranking function.

Experimental results show that our approach outperforms several state-of-the-art bottom-up (e.g., [13, 15, 16, 19, 20, 24, 68]) and top-down (e.g., [29, 44, 46]) approaches in visual saliency estimation. On the prevalent video eye-fixation dataset provided by Itti [22], our approach can reach an ROC score of 0.774. Some representative examples are illustrated in Fig. 4.7. From Fig. 4.7, we can see that our approach can effectively and accurately locate the entire salient objects in various scenes.

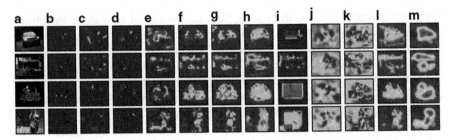

Fig. 4.7 Some representative results. Note that here the eye density maps are not convolved with a Gaussian kernel, which is a popular method to recover more positive samples for the evaluation. (**a**) Original frames; (**b**) eye fixation maps; (**c**) [24]; (**d**) [19]; (**e**) [20]; (**f**) [16]; (**g**) [13]; (**h**) [15]; (**i**) [68]; (**j**) [46]; (**k**) [29]; (**l**) [44]; (**m**) our approach

Fig. 4.8 Targets and distractors in different scenes can be best distinguished by different features. (**a**),(**b**) the "motion" feature; (**c**),(**d**) the "color" feature

4.3.3.4 Multi-Task Rank Learning for Visual Saliency Estimation in Video

Generally speaking, a unified ranking function derived with the proposed approach can obtain impressive results in some cases but meanwhile may suffer poor performance in other cases since they often construct a unified model for all scenes. Actually, the features that can best distinguish targets from distractors may vary remarkably in different scenes. In surveillance video, for instance, the motion features can be used to efficiently pop-out a car or a walking person (as shown in Fig. 4.8a, b); while to distinguish a red apple/flower from its surroundings, color contrasts should be used (as shown in Fig. 4.8c, d). In most cases, it is infeasible to pop-out the targets and suppress the distractors by using a fixed set of visual features. Therefore, it is necessary to construct scene-specific models that adaptively adopt different solutions for different scene categories.

Toward this end, we propose a multi-task rank learning approach for visual saliency estimation. In this approach, visual saliency estimation is also formulated as a pair-wise rank learning problem. However, this approach constructs multiple visual saliency models, each for a scene cluster, by learning and integrating the features that best distinguish targets from distractors in that cluster. We also propose

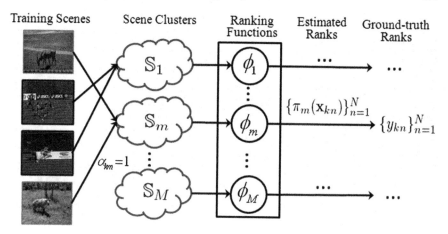

Fig. 4.9 The framework of our approach. In our approach, scenes with similar contents are grouped into the same cluster. For each cluster, a ranking function is optimized to give ranks for all subsets in a scene, while these estimated ranks are expected to approximate the ground-truth ranks

a multi-task learning algorithm to infer multiple saliency models simultaneously. Different from the traditional single-task learning approach, the multi-task learning approach can carry out multiple training tasks simultaneously with fewer training data per task. In this framework, the appropriate sharing of information across training tasks can be used to effectively improve the performance of each model.

The system framework of this approach is illustrated in Fig. 4.9. In this framework, the training scenes are grouped into M clusters and a ranking function is optimized for each cluster. Here the ranking function is optimized as in the previous approach (i.e., the same pair-wise losses, the same optimization strategies). However, several penalty terms are added into the optimization process to improve the performance of each of the M ranking functions, especially for the generalization ability. These penalty terms mainly consist of scene clustering penalty (to group scenes with similar contents into the same cluster), model diversity penalty (to improve the generalization ability of each ranking function), and model complexity penalty (to avoid over-complex model). By introducing these penalty terms, the training process can optimize M ranking functions simultaneously with an appropriate sharing of information across them. Therefore, the performance of this approach is much better than the other approaches (e.g., [13, 15, 16, 19, 20, 24, 30, 44, 46, 68]) and can reach a ROC score of 0.811 on the video eye-fixation dataset [22]. Some representative examples are illustrated in Fig. 4.10. From Fig. 4.10, we can see that our approach can adapt to various scenes and demonstrates a higher accuracy in locating the most salient targets.

Fig. 4.10 Some representative results of visual saliency models. (**a**) Original scenes; (**b**) Ground-truth saliency maps; (**c**) [24]; (**d**) [19]; (**e**) [20]; (**f**) [68]; (**g**) [15]; (**h**) [16]; (**i**) [13]; (**j**) [30]; (**k**) [44]; (**l**) [46]; (**m**) MTRL

4.4 Content Analysis for Genre-Specific Video

In this section, we will present two representative content analysis works in two types of video, i.e., sports video and surveillance video.

4.4.1 Sports Video Scene Analysis

Various innovative and original works have been applied and proposed in the field of sports video analysis. However, individual works focused on sophisticated method-ologies with particular sport types and there was a lack of scalable and holistic framework in this field. This section presents a solution for this issue and presents a systematic and generic approach which is experimented on relatively large-scale sports consortia.

4.4.1.1 A Generic Framework for Sports Video Analysis

A system overview from a holistic aspect is illustrated in Fig. 4.11, such that the input sports video is analyzed systematically using a generic and sequential frame-work. This is interpreted such that the result from a preceding process is input to the next process with a consistent and coherent fashion. The highlights of this frame-work include:

1. A generic foundation using domain-knowledge free local feature is developed to represent input sports videos. This method would fit the general framework in sports video analysis and provides an alternative solution to alleviate generality, scalability, and extension issues.

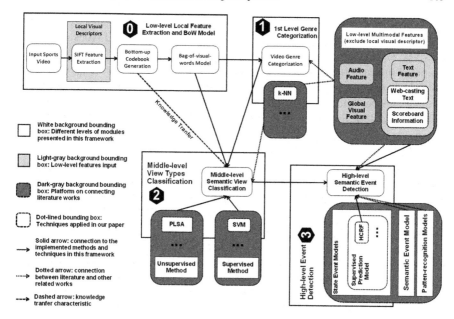

Fig. 4.11 A flowchart of the proposed generic framework, with one module of generic video representation and three task modules in sequence

2. A thorough and systematic structure starting from genre identification is presented, which was ignored in some related works by assuming the genre type as the prior knowledge.
3. A general platform is introduced, with one module of generic video representation and three task modules in sequence.

At the module 0, the low-level local feature utilization incorporating with the codebook generation and the BoW model provides an expandable groundwork for the semantic tasks of genre categorization, view classification and high-level event detection. In this structure, a homogenous process is first introduced for extracting domain-knowledge independent local descriptors. A BoW model is used to represent an input video by mapping its local descriptors to a codebook, which is generated from an innovative bottom-up parallel structure. The histogram based video representation is treated as sole input (no other feature models) to both the genre categorization and the view classification modules. Such a concise representation built from the BoW model benefits users in homogenously extract visual features and represent videos in a compact and collective form.

At the first module, the videos are categorized by its genre. Video genre nomenclature is used to describe the video type, which is defined as the highest level of granularity in video content representation. Since the video genre categorization task directly relies on low-level features, the proposed feature extraction of target video sequence is used in categorization. In large-scale videos, a successful identification of the genre serves as the first step before attempting higher level tasks. For instance, in sports event detection, an unknown "shooting" event is the target quest, which could be from the ball game or the shooting sport. By indiscriminately treating the entire dataset, this event will be searched through all types of sports. However, since sports like figure-skating and swimming have no "shooting" at all, the effort in searching this event at those non-relevant sports becomes infeasible. Instead of treating all data indifferently, a more efficient approach is to identify the genre of the query video first, and then deploy middle/high-level tasks consequently.

In the middle-level and the second module, semantic view types are classified using an unsupervised PLSA learning method to provide labels for input video frames. View describes an individual video frame by abstracting its overall content. It is treated as a bridge between low-level visual features and high-level semantic understanding. In addition, unsupervised learning saves a massive amount of human effort in processing large-scale data. Moreover, the supervised methods can also be implemented upon our proposed platform. Therefore, a SVM model is executed as the baseline for the comparison purpose.

Finally at the third module, a structured prediction HCRF model using labeled inputs is a natural fit to the system in detecting semantic events. This can be justified by the fact that a video event occupies various lengths along the temporal dimension. Thus, the state event model-based HCRF is suitable to deploy. Less comprehensive baseline methods such as the hidden Markov model and the conditional random field can also be applied in this platform.

4.4.1.2 High-Level Event Detection for Sports Video

Content-based video event detection is among the most popular quest for the high level semantic analysis. Different from video abstraction and summarization which target on any interesting events happening in a video rush, event detection is only constrained to a pre-defined request type, such as the third goal or the second penalty kick in a particular soccer match. In sports video, a consumers interest of events resides in the actual video contents, more than just the information delivered. On the other hand, sports videos also have a very strongly correlated temporal structure. In a way, such the structure can be interpreted as a sequence of video frames which have patterns and internal connections. This pattern existence is ubiquitous due to the nature of the sports, a competition where players learn from the standard in order to excel. Therefore, an intuitive approach is to find such patterns using certain representation and learn the temporal structure. Luckily, the PLSA approach provides such a labeled frame sequence and what we need is a clever technique on which portion of the video to analyze and what robust structured prediction model to use. Following, we will introduce a coarse-to-fine scheme and hidden conditional random field (HCRF) for event detection.

Before learning the tempo and patterns, a starting and entry point of an event needs to be seized. A two-stage coarse-to-fine event detection strategy is suitable for this scenario. The first stage is a rough event recognition and localization utilizing rich and accurate text-based information either from web-casting text or optical character recognition (OCR) techniques of the score-board update. In the second stage, precise video contents associated to the semantic event have been detected in terms of the event boundary detection and accuracy analysis. For example, Web-casting text for coarse stage event detection and video alignment was studied and analyzed such as replay scenes and various goal and shot scenes detection in soccer video [7, 61].

Since the proposed framework targets on the generic learning model that can be extended to large-scale, we propose a HCRF based structured prediction model utilizing previously classified views, and completing the generic approach. For example, the HCRF model can be used to detect the score event in basketball for exciting events and highlights. Such a HCRF technique belongs to the state event model defined in the related works. Therefore, the HCRF takes the labeled sequences as input in a natural and seamless fashion. On the other hand, the HCRF is a comprehensive model, which can be degraded to hidden Markov models (HMM) or conditional random fields (CRF) with certain constraints. The merits of HCRF comparing the other two models are its resilience and robustness with combination of both the hidden states and the Markov property relaxation.

There are several advantages of using the HCRF in large-scale datasets than HMM or CRF models. Firstly, HCRF relaxes the Markov property, which assumes that the future state only depends on the current state. In our generic framework, video frames are uniformly decimated and sampled, regardless of the temporal pace of video itself. In some cases, several consecutive frames have the same labeling while in other cases, different labels are assigned. Markov property based model such as HMM is appropriate for the former scenarios but not suitable for the latter ones, since the future state in HMM only cares about the current state label but not previous states. On the other hand, HCRF is flexible and takes surrounding states from both before and after the current state. Thus, HCRF is more robust for dealing with large-scale homogeneous process and uniform sampling with no prior knowledge. For instance, if a key frame immediate preceding the current stage is missed due to the uniform sampling. such an information loss could be compensated by including and summing up previous or later information without misclassifying the event. Secondly, HCRF has merit in its hidden states structure, which helps to relax the requirement of explicit observed states. This is also an advantage in dealing large-scale uniformly sampled video frames. It is because that in computation, the CRF model outputs individual result label (such as event or not event) per state and requires separate CRFs to present each possible event [62]. In HCRF, only one final result is presented in terms of multi-class events occurring probabilities. From the robustness point of view, a CRF model can be easily ruined by semantically unrelated frames due to the automatic uniform sampling. A multiclass HCRF on the other hand, can correct the error introduced by such unrelated frames using probability-based outputs [49]. Moreover, the HCRF is also appealing in allowing the use of not explicitly labeled training data with partial structure [49]. From literature, HCRF has been successfully used in gesture recognition [49] and phone classification [12].

Figure 4.12a illustrates a HCRF structure, in which a label $y \in Y$ of event type is predicted from an input X. This input consists of a sequence of vectors

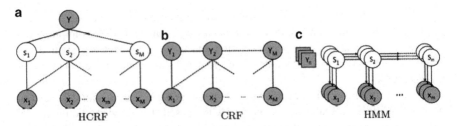

Fig. 4.12 Structured Prediction Models. (**a**): Hidden conditional random field (HCRF). (**b**): Conditional random field (CRF). (**c**): Hidden Markov Model (HMM)

$X = x_1, x_2, \ldots, x_m, \ldots, x_M$, with each x_m representing a local state observation along the HCRF structure. Different decision stages of aforementioned three structured prediction models are employed to detect an event. For the HMM, the query sequence will be tested and the highest likelihood of the HMM provides the final decision in event detection. On the other hand in the CRF model, since each state variable $Y(t)$ requires a label as Fig. 4.12b shows, a majority-rule voting scheme in which the most event labels along the Y sequence decide the event result. For the HCRF model depicted in Fig. 4.12a, a multiclass training process recognizing all classes at the same time is adopted. Therefore, a detected event with the highest probability is considered as the final result for the query sequence.

4.4.1.3 Experiments and Results

In the following, experimental results are presented to justify the properties of the proposed generic framework, specifically using a relatively large-scale video collection including 23 genres with a total of 145 h gathered by the authors, named as 23-sports dataset. All the video clips have the same length of 167 s with a total of 500 uniformly sampled frames at a sampling rate of 3 frames per second. This dataset is composed of 3122 clips. In training, 1,198 clips are used, in which a subset of 46 clips (2 clips per sport) are used in codebook generation with a total of 3,112,341 SIFT points. In testing, the other 1,924 clips are selected.

In this experiment, the task on basketball score event detection is investigated by employing this labeled video sequence. Two-staged coarse-to-fine scheme is adopted with firstly detecting scoreboard information change introduced by [39]. By adopting this technique, an entry point of an interesting event is located. However, this coarse detection only provides a static frame based rough estimation as an entry point. Since scoreboard information not only appears in score events, but also in time-out events or intermission events, individual frame based detection without temporal structured information cannot provide robust and satisfactory result. Therefore, a fine tuning process in finalizing detection is adopted to ensure that the query video truly conveys the score event as its semantic theme. The proposed HCRF model is deployed as such process after the first stage coarse detection. Experimental results of using this HCRF model are compared with CRF and HMM baselines.

Two video groups consisting of four matches are utilized, which are defined as (a) Dataset A: using two NBA games for training, and using another two Olympic Games for testing; (b) Database B: using one NBA game for training, and using another NBA game for testing. Frame-based views from the PLSA model and the SVM model are applied to Dataset A and B. Therefore, four combinations of view labels and datasets are defined as PLSA+A, PLSA+B, SVM+A, and SVM+B. Each video clip used in both training and testing is automatically decimated and consists of 500 uniformly sampled frames. We use a window size $N = 20$, with a window N sliding every 10 frames.

The number of approximated events detected after the first stage is given in Table 4.2. The precision and recall of the coarse stage basketball score detection are 92.03% and 86.19% respectively. In the second stage, the proposed HCRF-based model and state of the art HMM and CRF models are evaluated and compared. The advantage of HCRF over HMM is its relaxation on the Markov property that the current state St can be inferred from both current observation as well as surrounding observations. As shown in Table 4.3, the HCRF has better performance than the CRF for the same ω values, while both models outperform the HMM baseline. When using different ω values for both CRF and HCRF, $\omega = 1$ provides better results than $\omega = 0$, in which neighboring information assists in a better decision making. However, when $\omega = 2$ is used for HCRF, the performance has been dropped for all cases comparing with $\omega = 1$. This can be viewed as an over-fitting issue, in which adding more surrounding information limits the structured prediction ability. In summary, the proposed HCRF based model with parameter $\omega = 1$ outperforms both CRF and HMM models. The best results are obtained at 93.08% and 92.31% by taking SVM and PLSA based input labels, respectively.

Table 4.2 Precision and recall results of basketball score events detection at the first (coarse) stage

Correctly detected score (True positive)	Detected score (Obtained result)	Correct total score (Obtained result)	Precision (%)	Recall (%)
231	251	268	92.03	86.19

Table 4.3 Performance comparison on score event detection in basketball. Dataset A: NBA matches as training, Olympic matches as testing. Dataset B: NBA matches for both training and testing

	Accuracy			
	Dataset A (NBA/Olympics)		Dataset B (NBA/NBA)	
	SVM+A(%)	PLSA+A(%)	SVM+B(%)	PLAS+B(%)
HMM $\omega = 0$	78.28	75.29	87.50	85.94
CRF $\omega = 0$	78.16	74.57	87.43	86.52
CRF $\omega = 1$	79.52	76.82	88.52	87.89
HCRF $\omega = 0$	80.93	75.53	90.00	90.77
HCRF $\omega = 1$	83.26	80.24	93.08	92.31
HCRF $\omega = 2$	82.09	77.88	91.46	91.77

4.4.2 Surveillance Video Event Detection

In this section, we present a system for TRECVID'09 surveillance event detection tasks. Two categories of events are detected in this system: (1) single-actor events (i.e., *PersonRuns* and *ElevatorNoEntry*) irrespective of any interaction between individuals, and (2) pair-activity events (i.e., *PeopleMeet*, *PeopleSplitUp*, and *Embrace*) that involves more than one individual. Figure 4.13 shows the framework of this system.

The system consists of three major stages, i.e., preprocessing, event classification, and post-processing. The preprocessing involves view classification, background subtraction, head-shoulder detection, human body detection and object tracking. Event classification fuses One-vs.-All SVM and automata-based classifiers to identify single-actor and pair-activity events in an ensemble way. To reduce false alarms, events merging and post processing based on prior knowledge are applied to refine system detection outputs.

4.4.2.1 Pair-Activity Events

Pair-activity events involve the interaction of at least two persons. This is dealt with as a classification problem. For pair-activity events, the events of *PeopleMeet*, *PeopleSplitUp* and *Embrace* are first treated as one category and One-vs.-All SVM is used to classify them from the others. Each kind of three events is identified by object motion patterns.

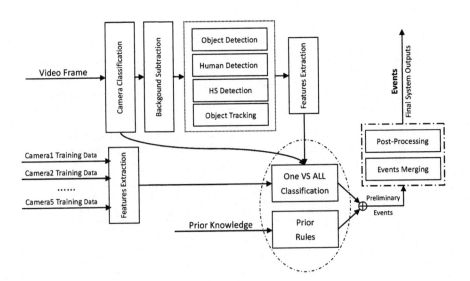

Fig. 4.13 A system designed for TRECVID'09 surveillance event detection tasks

Given two detected people, their distance, coexisting duration, and motion direction's correlation are combined to form a feature vector, which are extracted in a sliding window of twelve consecutive frames. One-vs.-All SVM is trained to classify this general category of events. To distinguish "*Embrace*" or "*PeopleMeet*", a backward search is applied to locate the beginning of an event. In contrast, forward search is used to detect "*PeopleSplitUp*".

Finally, the results are improved with post-processing in which a set of heuristic rules are used. For instance, if two peoples' distance at the end of an event is greater than a threshold for "*PeopleMeet*" and "*Embrace*", or their distance at the beginning of an event is greater than a threshold for "*PeopleSplitUp*", those would be considered as a false alarm.

4.4.2.2   Single-Actor Events

Speed and direction of movements are key characteristics of "*PersonRuns*". It is observed that a running person have a larger velocity than others, and the motion direction would not change dramatically. According to the feature statistics, the constraints of object position and motion direction can be used in a SVM classifier to identify *PersonRuns*. In the camera setting of TRECVID dataset, running people always move from left-bottom to top in the view of camera one. So, by using the post-processing step we may remove many false alarms caused by tracking drifting.

ElevatorNoEntry is defined as "elevator doors open with a person waiting in front of them, but the person does not get in before the doors close". As illustrated in Fig. 4.14, an automaton is adopted to model the detection process of *ElevatorNoEntry*. As there is no elevator in the view of cameras one, two, and five over TRECVID'09, the automaton is executed only in the views of camera three or four.

As the elevator's position is fixed, the system can easily locate the elevator. When the elevator door is closed, the elevator region is labeled as background. And when

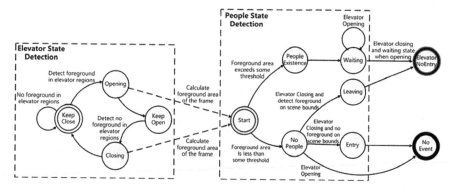

Fig. 4.14 An automaton for detecting *ElevatorNoEntry*

Table 4.4 Results of the system in the TRECVID-ED 2009 formal evaluation

Event	#Ref	#Sys	#CorDet	#FA	#Miss	Act. DCR
PeopleMeet	449	125	7	118	442	1.023
PeopleSplitUp	187	198	7	191	180	1.025
Embrace	175	80	1	79	174	1.02
ElevatorNoEntry	3	4	2	2	1	0.334

the door is moving, the elevator region is detected as foreground. Thus, each elevator's states (open or closed) can be identified by using background subtraction. The elevator's opening and closing moments are thus detected and recorded. The foreground area size is related to the number of persons in front of an elevator. Given an incoming frame, the *ElevatorNoEntry* event can be detected according to the elevators' states and the size of foreground area. Moreover, the area ratio of detected foreground regions before and after the detected elevator open-and-close action can be computed to tell whether the number of persons around an elevator has changed. When the ratio is less than a threshold, it is probable that some people enter the elevator, and the frame interval is labeled as a potential event of *ElevatorEntry*.

4.4.2.3 Evaluation Results in TRECVID 2009

This system is evaluated over the surveillance event detection task in TRECVID 2009. Three runs were submitted, by using different human detection and tracking modules. According to the comparative results in the TRECVID-ED formal evaluation, the experimental results are promising. Among all submissions for the formal evaluation, four detection results (e.g., *PeopleMeet, PeopleSplitUp, Embrace*, and *ElevatorNoEntry*) ranked at the first place (Table 4.4).

However, there are some problems yet. Regarding the results, the false alarms have been reduced greatly by effective post-processing. Unfortunately, much correct detection is wiped off at the same time. In other words, the system recall is too low. Furthermore, the system should make a good tradeoff between false alarms and system recall.

4.5 Summary

This chapter presents a tutorial on the problems and solutions of video scene analysis from the perspectives of the learning components and tasks. Two major categories of the state-of-the-art tasks are discussed in this chapter, based on their application setup and learning targets: generic methods and genre-specific analysis techniques. Many of the works discussed in this chapter are deemed by the authors as good representatives of existing learning-based video scene analysis technologies. Clearly, this chapter is only meant to capture the landscape of the field that is still young and still evolving. For a long-term perspective, video scene analysis is an interesting issue that currently requires for a lot more research efforts.

Acknowledgement The work is supported by grants from the Chinese National Natural Science Foundation under contract No. 60973055 and No. 61035001, and National Basic Research Program of China under contract No. 2009CB320906.

References

1. S. Aksoy, K. Koperski, C. Tusk, G. Marchisio, and J.C. Tilton, "Learning Bayesian classifiers for scene classification with a visual grammar," IEEE Trans. Geoscience and Remote Sensing, vol. 43, no. 3, pp. 581-589, 2005.
2. Y. Altun, I. Tsochantaridis, and T. Hofman, "Hidden Markov support vector machines," in Proc. IEEE Int. Conf. Mechine Learning, 2003, pp. 3-10.
3. K. Barnard, P. Duygulu, N. de Freitas, D. Forsyth, D. Blei, and M. I. Jordan, "Matching words and pictures," J. Machine Learning Research, vol 3, pp. 1107-1135, 2003.
4. S. Boyd and L. Vandenberghe. Convex Optimization. Cambridge University Press, 2004.
5. N. D. Bruce and J. K. Tsotsos. Saliency based on information maximization. In Advances in neural information processing systems, pp. 155-162, 2006.
6. M. Cerf, J. Harel, W. Einhauser, and C. Koch, Predicting human gaze using low-level saliency combined with face detection, in Advances in Neural Information Processing Systems, 2008, pp. 241-248.
7. Dai, J., Duan, L., Tong, X., Xu, C., Tian, Q., Lu, H., and Jin, J. 2005. Replay scene classification in soccer video using web broadcast text. In Proc. IEEE ICME. 1098-1101.
8. L. Duan, I.W. Tsang, D. Xu, and S.J. Maybank, "Domain transfer SVM for video concept detection," in Proc. IEEE Int. Conf. Computer Vision and Pattern Recognition, 2009, pp. 1-8.
9. S. Ebadollahi, L. Xie, S.-F., Chang, and J.R. Smith, "Visual event detection using multidimensional concept dynamics," in Proc. IEEE Int. Conf. Multimedia and Expo, 2006, pp. 881-884.
10. C. Frith. The top in top-down attention. In Neurobiology of attention (pp. 105-108), 2005.
11. Wen Gao, Yonghong Tian, Tiejun Huang, Qiang Yang. Vlogging: A Survey of Video Blogging Technology on the Web. ACM Computing Survey, 2(4), Jun. 2010.
12. Gunawardana, A., Mahajan, M., Acero, A., and Platt, J. 2005. Hidden conditional random fields for phone classification. In Proc. Interspeech. 1117-1120.
13. C. Guo, Q. Ma, and L. Zhang, Spatio-temporal saliency detection using phase spectrum of quaternion fourier transform, in IEEE Conference on Computer Vision and Pattern Recognition, 2008.
14. J. S. Hare, P. H. Lewis, P. G. B. Enser and C. J. Sandom, "Mind the Gap: Another look at the problem of the semantic gap in image retrieval," Multimedia Content Analysis, Management and Retrieval 2006, vol. 6073, No. 1, 2006, San Jose, CA, USA.
15. J. Harel, C. Koch, and P. Perona, Graph-based visual saliency, in Advances in Neural Information Processing Systems, 2007, pp. 545-552.
16. X. Hou and L. Zhang, Saliency detection: A spectral residual approach, in IEEE Conference on Computer Vision and Pattern Recognition, 2007.
17. H. Hsu, L. Kennedy, and S. F. Chang, "Video search reranking through random walk over document-level context graph," in Proc. ACM Multimedia, 2007, pp. 971-980.
18. Y. Hu, D. Rajan, and L.-T. Chia, Robust subspace analysis for detecting visual attention regions in images, in ACM International Conference on Multimedia, 2005, pp. 716-724.
19. L. Itti and C. Koch, Computational modeling of visual attention, Nature Review Neuroscience, vol. 2, no. 3, pp. 194-203, 2001.

20. L. Itti and P. Baldi, A principled approach to detecting surprising events in video, in Proc. IEEE Conference on Computer Vision and Pattern Recognition, 2005, pp. 631-637.
21. L. Itti, G. Rees, and J. Tsotsos. Neurobiology of attention. San Diego: Elsevier, 2005
22. L. Itti, Crcns data sharing: Eye movements during free-viewing of natural videos, in Collaborative Research in Computational Neuroscience Annual Meeting, 2008.
23. L. Itti and C. Koch. Feature combination strategies for saliency-based visual attention systems. Journal of Electronic Imaging, 10(1), 161-169, 2001.
24. L. Itti, C. Koch, and E. Niebur, A model of saliency-based visual attention for rapid scene analysis, IEEE Transactions on Pattern Analysis and Machine Intelligence, vol. 20, no. 11, pp. 1254-1259, 1998.
25. W. Jiang, S. F. Chang, and A. Loui, "Context-based concept fusion with boosted conditional random Fields," in Proc. Int. Conf. on Acoustics, Speech, and Signal Processing, 2007, pp. 949-952.
26. Shuqiang Jiang, Yonghong Tian, Qingming Huang, Tiejun Huang, Wen Gao. Content-Based Video Semantic Analysis. Book Chapter in Semantic Mining Technologies for Multimedia Databases (Edited by Tao, Xu, and Li), IGI Global, 2009.
27. Y. G. Jiang, J. Wang, S. F. Chang, C. W. Ngo, "Domain adaptive semantic diffusion for large scale context-based video annotation," in Proc. IEEE Int. Conf. Computer Vision, 2009, pp. 1-8.
28. L. Kennedy, and S. F. Chang, "A reranking approach for context-based concept fusion in video indexing and retrieval," in Proc. IEEE Int. Conf. on Image and Video Retrieval, 2007, pp. 333-340.
29. W. Kienzle, F. A.Wichmann, B. Scholkopf, and M. O. Franz, A nonparametric approach to bottom-up visual saliency, in Advances in Neural Information Processing Systems, 2007, pp. 689-696.
30. W. Kienzle, B. Scholkopf, F. A. Wichmann, and M. O. Franz, How to find interesting locations in video: a spatiotemporal interest point detector learned from human eye movements, in 29th DAGM Symposium, 2007, pp. 405-414.
31. M. Li, Y. T. Zheng, S. X. Lin, Y. D. Zhang, T.-S. Chua, Multimedia evidence fusion for video concept detection via OWA operator, in Proc. Advances in Multimedia Modeling, pp. 208-216, 2009.
32. H. Liu, S. Jiang, Q. Huang, C. Xu, and W. Gao, Region-based visual attention analysis with its application in image browsing on small displays, in ACM International Conference on Multimedia, 2007, pp. 305-308.
33. T. Liu, J. Sun, N.-N. Zheng, X. Tang, and H.-Y. Shum, Learning to detect a salient object, in IEEE Conference on Computer Vision and Pattern Recognition, 2007.
34. T. Liu, N. Zheng, W. Ding, and Z. Yuan, Video attention: Learning to detect a salient object sequence, in IEEE International Conference on Pattern Recognition, 2008.
35. Y. Liu, F. Wu, Y. Zhuang, J. Xiao, "Active post-refined multimodality video semantic concept detection with tensor representation," in Proc. ACM Multimedia, 2008, pp. 91-100.
36. K. H. Liu, M. F. Weng, C. Y. Tseng, Y. Y. Chuang, and M. S. Chen, "Association and temporal rule mining for post-processing of semantic concept detection in video," IEEE Trans. Multimedia, 2008, pp. 240-251.
37. Y.-F. Ma, X.-S. Hua, L. Lu, and H.-J. Zhang, A generic framework of user attention model and its application in video summarization, IEEE Transactions on Multimedia, vol. 7, no. 5, pp. 907-919, 2005.
38. S. Marat, T. H. Phuoc, L. Granjon, N. Guyader, D. Pellerin, and A. Guerin-Dugue, Modelling spatio-temporal saliency to predict gaze direction for short videos, International Journal of Computer Vision, vol. 82, no. 3, pp. 231-243, 2009.
39. G. Miao, G. Zhu, S. Jiang, Q. Huang, C. Xu, and W. Gao, A Real-Time Score Detection and Recognition Approach for Broadcast Basketball Video. In Proc. IEEE Int. Conf. Multimedia and Expo, 2007, pp. 1691-1694.
40. F. Monay and D. Gatica-Perez, "Modeling semantic aspects for cross-media image indexing," IEEE Trans. Pattern Anal. Mach. Intell., vol. 29, no. 10, pp. 1802-1917, Oct. 2007.

41. M. R. Naphade, I. Kozintsev, and T. Huang, "Factor graph framework for semantic video indexing," IEEE Trans. Circuits and Systems for Video Technology, 2002, pp. 40-52.
42. M. R. Naphade, "On supervision and statistical learning for semantic multimedia analysis," Journal of Visual Communication and Image Representation, vol. 15, no. 3, pp. 348-369, Sep. 2004.
43. A. Natsev, A. Haubold, J. Tesic, L. Xie, R. Yan, "Semantic concept-based query expansion and re-ranking for multimedia retrieval," in Proc. ACM Multimedia, 2007, pp. 991-1000.
44. V. Navalpakkam and L. Itti, Search goal tunes visual features optimally, Neuron, vol. 53, pp. 605-617, 2007.
45. T. N. Pappas, J.Q. Chen, D. Depalov, "Perceptually based techniques for image segmentation and semantic classification," IEEE Communications Magazine, vol. 45, no. 1, pp. 44-51, Jan. 2007.
46. R. J. Peters and L. Itti, Beyond bottom-up: Incorporating task-dependent influences into a computational model of spatial attention, in IEEE CVPR, 2007.
47. R. J. Peters and L. Itti. Congruence between model and human attention reveals unique signatures of critical visual events. In Advances in neural information processing systems (pp. 1145-1152), 2007.
48. G. J. Qi, X. S. Hua, Y. Rui, J. Tang, T. Mei, and H. J. Zhang, "Correlative multi-label video annotation," in Proc. ACM Multimedia, 2007, pp. 17-26.
49. Quattoni, A.,Wang, S., Morency, L., Collins, M., Darrell, T., and Csail, M. 2007. Hidden state conditional random fields. IEEE Transactions on Pattern Analysis and Machine Intelligence 29, 10, 1848-1852.
50. A.W.M. Smeulders, M. Worring, S. Santini, A. Gupta, and R. Jain, "Content-based image retrieval at the end of the early years," IEEE Trans. Pattern Anal. Mach Intell., vol. 22, no.12, pp. 1349-1380, Dec. 2000.
51. J. R. Smith, M. Naphade, and A. Natsev, "Multimedia semantic indexing using model vectors," in Proc. IEEE Int. Conf. Multimedia and Expo, 2003, pp. 445-448.
52. C. G. M. Snoek, M. Worring, J.C. Gemert, J.-M. Geusebroek, and A.W.M. Smeulers, "The challenge problem for automated detection of 101 semantic concepts in multimedia," in Proc. ACM Multimedia, 2006, pp. 421-430.
53. E. Spyrou and Y. Avrithis, "Detection of High-Level Concepts in Multimedia," Encyclopedia of Multimedia, 2nd Edition, Springer 2008.
54. A. M. Treisman and G. Gelade, A feature-integration theory of attention, Cognitive Psychology, vol. 12, no. 1, pp. 97-136, 1980.
55. I. Tsochantaridis, T. Hofmann, T. Joachims, and Y. Altun, "Support vector machine learning for interdependent and structured output spaces," in Proc. IEEE Int. Conf. Machine Learning, 2004, pp. 823-830.
56. D. Walther and C. Koch, Modeling attention to salient proto-objects, Neural Networks, vol. 19, no. 9, pp. 1395-1407, 2006.
57. T. Wang, J. Li, Q. Diao, W. Hu, Y. Zhang, and C. Dulong, "Semantic event detection using conditional random fields," in Proc. IEEE Int. Conf. Computer Vision and Pattern Recognition Workshop, 2006.
58. M. Weng, Y. Chuang, "Multi-cue fusion for semantic video indexing," in Proc. ACM Multimedia, 2008, pp. 71-80.
59. L. Xie and S. F. Chang, "Structural analysis of soccer video with hidden markov models," in Proc. Int. Conf. on Acoustics, Speech, and Signal Processing, 2002, pp. 767-775.
60. Xiong, Z. Y., Zhou, X. S., Tian, Q., Rui, Y., and Huang, T. S. Semantic retrieval of video: Review of research on video retrieval in meetings, movies and broadcast news, and sports. IEEE Signal Processing Magazine 18, 3, 18-27, 2006.
61. Xu, C., Wang, J., Wan, K., Li, Y., and Duan, L. 2006. Live sports event detection based on broadcast video and web-casting text. In Proc. ACM MM. 230.
62. Xu, C., Zhang, Y., Zhu, G., Rui, Y., Lu, H., and Huang, Q. 2008. Using webcast text for semantic event detection in broadcast sports video. IEEE Transactions on Multimedia 10, 7, 1342-1355.

63. R. Yan, M. Y. Chen, and A. Hauptmann, "Mining relationship between video concepts using probabilistic graphical models," in Proc. IEEE Int. Conf. Multimedia and Expo, 2006, pp. 301-304.
64. J. Yang and A. G. Hauptmann, "Exploring temporal consistency for video analysis and retrieval," in Proc. 8th ACM SIGMM Int. Workshop on Multimedia Information Retrieval, 2006, pp. 33-46.
65. J. Yang, R. Yan, A. Hauptmann, "Cross-domain video concept detection using adaptive svms," in Proc. ACM Multimedia, 2007, pp. 188-297.
66. Yang Yang, Jingen Liu, Mubarak Shah, Video Scene Understanding Using Multi-scale Analysis, Proc. 12th Int'l Conf. Computer Vision, 1669-1676, 2009.
67. Zheng-Jun Zha, Linjun Yang, Tao Mei, Meng Wang, Zengfu Wang, Tat-Seng Chua, Xian-Sheng Hua. Visual query suggestion: Towards capturing user intent in internet image search. ACM Transactions on Multimedia Computing, Communications, and Applications, 6(3), Article 13, August 2010.
68. Y. Zhai and M. Shah, Visual attention detection in video sequences using spatiotemporal cues, in ACM International Conference on Multimedia, 2006, pp. 815-824.
69. H. Zhang, A. C. Berg. M. Maire, and J. Malik, "Svm-knn: Discriminative nearest neighbor classification for visual category recognition," Proc. IEEE Conf. CVPR, pp. 2126-2136, 2006.

Chapter 5
Multiview Image Segmentation and Video Tracking

King Ngi Ngan and Qian Zhang

Abstract Image segmentation and video tracking (ISVT) is a necessary and important preliminary step in many high-level vision tasks such as activity recognition, rendering and modeling, and scene analysis. Comparing with the monoview ISVT, multiview ISVT is capable of characterizing the visual object and dynamic scene with three-dimensional (3D) interpretation, which prevails over the traditional two-dimensional (2D) representation. In this chapter, we categorize and review the representative and state-of-the-art approaches in multiview image segmentation and video tracking. Additionally, our proposed depth-based segmentation in the initial frame and feature-based tracking algorithms from multiview video for both separated and overlapped human objects are discussed respectively, following the ensuring experimental results to demonstrate the algorithms' superior performance.

5.1 Introduction

In the recent decades, image segmentation and video tracking (ISVT) has become an active research topic in image processing, computer vision and computer graphics, leading to the significant breakthroughs on the development of its theories and technologies. Most of content-based applications are more interested in accessing and manipulating meaningful objects instead of the whole scene, which makes the object-based segmentation and tracking in demand. Object-based ISVT aims to segment the image or video frame into a few semantic object-of-interests (OOIs) that are described as distinct spatial entities, and track the trajectory of these entities across the temporal sequence. Object-based representation can be achieved by decomposing the image or frames into meaningful objects, and visual information can be further viewed and edited once the silhouette of the objects are available. Thus,

K.N. Ngan (✉)
Department of Electronic Engineering, The Chinese University of Hong Kong,
Hong Kong, China
e-mail: knngan@ee.cuhk.edu.hk

K.N. Ngan and H. Li (eds.), *Video Segmentation and Its Applications*,
DOI 10.1007/978-1-4419-9482-0_5, © Springer Science+Business Media, LLC 2011

robust and accurate object segmentation and tracking from image and video has turned out to be the crucial prerequisite to facilitate various computer vision and advanced multimedia tasks such as the object-based video coding in MPEG4 standard [38], video object cut for pasting [23], face segmentation in videotelephony [4], 3D object modeling by joint segmentation [36], motion segmentation for scene understanding [42], human body configuration recovering and recognition [28], and video object extraction in surveillance system [16].

According to the type of the source data, object-based segmentation can be categorized into image segmentation or video segmentation. Even though the video sequence is composed of a collection of images or frames, video segmentation is different from segmentation of the single image. Video segmentation may incorporate image segmentation technique to segment each frame into lots of homogenous regions. However, temporal coherence constraint in the sequences results in the difference between video segmentation and the segmentation of series of its single frame. Temporal coherence constraint addresses strong correlation of segmentations overtime, but the results can be quite unstable if segmenting them independently using image segmentation algorithm.

Based on different camera configurations, OOIs can be segmented from single or multiple views of image/video. Comparing with the intensive and well-studied work on monoview ISVT, multiview ISVT has not been attracted much attentions due to the limitation of data acquisition technology and the difficulty to segment all the images simultaneously in realtime. However, multiview image/video (MVI/V) capturing the real-world environment from arbitrary viewpoints are capable of describing dynamic scene from different perspective and can provide the observer more vivid and extensive viewing experience than the monoview image/video, resulting in more realistic and exciting visual effect. Furthermore, depth information reconstructed from multiview data assists in recovering scene structures, and characterizing the visual objects using 3D model, which is more efficient and desirable than the conventional 2D representation. Whereas, the major obstacles of multiview image segmentation and video tracking are the expensive acquisitions, extremely large amount of data and intensive computational load.

With the recent advance in the multimedia processing and emergence of new generation digital devices, multiview capturing system with sparse or dense camera array [7] can be built with ease, which motivates the development of multiview techniques and its related applications. By producing several image sequences taken from different viewpoints of the real environment, MVI/V can generate more vivid and accurate information about the scene structure, resulting in the 3D feeling as the available depth information. Multiview ISVT provides the capability of describing dynamic object with multiple angles, and thus has widely been applied into numerous 3D functionalities, for example, image-based rendering, virtual reality and 3D security. Manually separating the object from background from MVI/V is known to be a tedious and time-consuming work due to the associated great quantities of data. Semi-supervised approaches can release the computational cost to certain extend by introducing human's interventions. However, the segmentation results are highly dependent on these interactions. Even though fully automatic method is the most efficient way for MVI/V segmentation, this still remains a challenging problem in the research community because of the insufficient accuracy and robustness.

In this chapter, we focus on the techniques of object-based image segmentation and video tracking from MVI/V. The rest of the chapter is divided into two parts: multiview image segmentation and multiview video tracking. In the reminder, image segmentation from multiview images is discussed in Sect. 5.2, with compressive review on the current algorithms and description of the proposed method. Section 5.3 addresses the overview on the existing algorithms following our proposed method on the topic of video segmentation and tracking from multiview video. Finally, conclusions are drawn in Sect. 5.4.

5.2 Multiview Image Segmentation

A series of multiview images (MVIs) can be either simultaneously captured by multiple cameras, or more commonly and economically, collected by a single camera at different viewpoints and time instances. According to the visual content, multiview image segmentation can be grouped in to region-based segmentation and object-based segmentation. The categorical overview of multiview image segmentation approaches is shown in Fig. 5.1. Region-based segmentation aims to cluster the perceptually similar pixels in the image into homogenous regions, while object-based segmentation tries to extract the meaningful object and separate foreground from background. Region-based segmentation focuses on the interpreting and understanding of the whole scene which is represented by semantically and geometrically consistent partitions as shown in Fig. 5.2b. On the contrary, object-based segmentation pays more attention to access and manipulate the OOIs, and the extracted objects are highlighted in the foreground mask as shown in Fig. 5.2d.

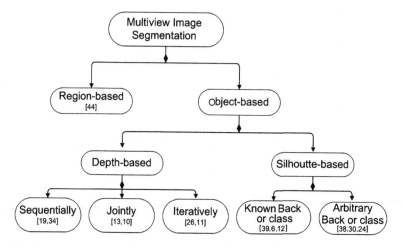

Fig. 5.1 Overview of multiview image segmentation approaches

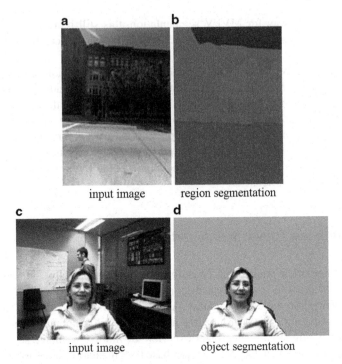

Fig. 5.2 Region-based and object-based multiview image segmentation

5.2.1 Region-Based Multiview Image Segmentation

The performance of understanding the semantic content of image and dynamic scene can be improved by the region-based segmentation and classification from multiple view images. Xiao and Quan [44] proposed a simple and powerful multiview segmentation framework, on the Google Maps Street View images captured by a camera mounted on a car driving along the street. A Markov Random Field is defined with associated graph for the multiple images in the same sequence, where a node in the graph corresponds with a super-pixel by over-segmentation. The extracted 2D image-based appearance and position features, as well as the 3D geometric features are collected to learn the AdaBoost classifiers for each class label and define the unary potential function in the Gibbs energy. Color difference in the same image and dense corresponds across different views are utilized to enforce smoothness and consistency. The segmentation results apply to the recognition of street view scene containing several semantic classes such as ground, building, sky, vehicle and person.

Decomposing the image into multiclass regions is a challenging task due to various visual concepts involved. In certain applications, the users are more interested in accessing a specific object rather than the scene, which makes the object-based segmentation in demand. A comprehensive review on the object-based multiview image segmentation and our proposed algorithm will discuss in the following sections.

5.2.2 Object-Based Multiview Image Segmentation

Localizing and extracting the OOIs in MVIs is the objective for object-based multiview image segmentation. Based on the different methodologies involved, the existing algorithms of object-based multiview image segmentation can be grouped into depth-based segmentation and silhouette-based segmentation.

5.2.2.1 Depth-Based Segmentation

Depth information reconstructed from MVIs usually serves as a valuable source in various related techniques such as 3D reconstruction [18], image-based rendering [32], freeview video generation [31], MVI/V compression [45] and virtual view synthesis [46]. Comparing with the 2D analysis and processing, the recovered depth information from the geometric relationship of MVIs assists in understanding and visualizing the 3D world in more efficient way. Accurate object segmentation in the clutter scene and complicated scenario is almost impossible or error-prone without any semantic knowledge about the scene or only relying on the 2D information (color, texture, and spatial location) from single-view images, since the semantic object is not always homogenous with these low-level features. By assuming that object locates in the different depth layer in the 3D scene and the depth value over one object forms smooth and consistent distributions, semantic objects can be extracted with known depth and segmentation performance using 2D features can be improved. However, object segmentation only exploiting the depth data is problematic due to the inaccuracy of the depth reconstruction resulting from the inherent difficulties of stereo matching such as the lack of textures and occlusion. Thus, to obtain more precise and robust segmentation for object-level manipulation, intelligent fusion of depth with other features should be taken into account.

Depth reconstruction and multiview segmentation is generally addressed in the sequential, joint or iterative fashion in a number of literatures. The most straightforward way for depth-based segmentation is to perform depth estimation beforehand, and then incorporate the depth information into the segmentation framework. Kolmogorov et al. in [19] described models and algorithms for bilayer segmentation of stereoscopic frames. Stereo disparity is obtained by dynamic programming in Layered Dynamic Programming algorithm, and stereo match likelihood is then probabilistically fused with contrast-sensitive color model to segment each frame by ternary graph cut. Good quality segmentation of temporal sequence can be achieved by marginalizing explicit temporal consistency in the realtime system. To automatically align the panoramic images and segment building from multiview city-scale street view, a fast and accurate method for multiview alignment and segmentation is proposed in [34]. Buildings in each panoramic image are labeled using graph cut image segmentation based on color and orientation features. The mistakes in single-view segmentation are corrected by aggregating the results over multiple views with the help of available depth, in which a much better segmentation can be obtained.

To prevent the propagation of error from stereo estimation to foreground extraction in the sequential approaches, depth reconstruction and object segmentation problems are simultaneously solved by joint optimization. For the challenging outdoor environments analysis with moving cameras, for example, rugby and soccer scenes [13], multiview scene reconstruction and segmentation are dealt with by joint graph-cut optimization. Segmentation and depth labeling field are formulated into the unified energy function, which involves color and contrast term for segmentation, as well as the match and smoothness term for reconstruction. Joint segmentation and reconstruction enables the high-quality scene representation of the sport scene. By exploiting strong interdependency between 3D reconstruction and foreground extraction, Golducke et al. [10] proposed a flexible and homogenous approach to simultaneous depth estimation and background subtraction in a multiview setting, assisted by a static background image with known depth for each camera. The results of depth reconstruction and background separations algorithm is obtained as minimization of energy functional, to generation a dense depth map and foreground map.

The iterative depth-based segmentation receives the segmentation feedback from current estimation to improve the depth reconstruction and vice verse. In order to create the intermediate synthesized view using depth and segmentation information, an iterative algorithm is developed in [26] which continuously performance the disparity estimation and the image segmentation in the iterative circle, and improve the result of each other. In [11], the estimated depth map and segmentation mask are iteratively computed using an Expectation-Maximization (EM) algorithm.

5.2.2.2 Silhouette-Based Segmentation

Object segmentation and reconstruction of 3D shapes are highly related topics in the field of computer vision and graphics. On the one hand, the acceptable segmentation of object with accurate silhouette from considerate amount of MVIs is required to accomplish the 3D object reconstruction, namely *Shape-from-silhouette* [5, 22] to combine the multiple silhouettes of same object from different viewpoints as source of shape information to reconstruct the 3D models. On the other hand, the reconstructed *visual hull* [20] from silhouette images, which approximately represents the geometry of object by linking the object shape in MVIs, is capable of refining the segmentation by projecting the visual hull onto the image plane and exploiting the *silhouette coherence* across MVIs.

Silhouette-based object segmentation from MVIs has been addressed in the recently extensive literatures. Tsai et al. [39] proposed a semiautomatic MVIs segmentation algorithm for 3D modeling by integrating with visual hull reconstruction. In the segmentation process, the automatic segmentation initialization is first carried out by graph-cut based image segmentation activated by trimap labelling. Then, it asks for user's interaction to choose a subset of the segmentation results with satisfactory quality for 3D reconstruction using volumetric graph cut and learning shape priors. At last, those discontent segmentations will be refined with the help of 3D model projection and the learned shape priors to propagate the successful segmentation and rectify the segmentation errors. An automatic algorithm in [6] dedicated to obtain the 3D segmentation of rigid object using volumetric graph-cut. Camera fixation constraint is adopted to initialize the OOI and the color model. Iterative refinement of silhouette extraction and visual hull estimation is performed by volumetric graph-cut optimization, which ensures that the resulting silhouettes by propagating the computed visual hull back to the individual view are consistent with one another at every iteration. Grauman et al. [12] presented a Bayesian approach to visual hull reconstruction using image-based representation of extracted silhouette from pedestrian images. The basic background subtraction results in rough segmentation corrupted by noise of each viewpoint in the simple color background. The visual hull of pedestrian is reconstructed by PPCA-based Bayesian model from problematic silhouettes. The used class-specific prior in visual hull reconstruction reduces the effect of segmentation errors in the silhouette extraction process.

The multiview segmentation algorithms using silhouette and visual hull as mentioned above are developed for equi-tilt set or tunable sequence relying on the known background, or simple background for a specific object class. For the silhouette extraction with arbitrarily unknown background, object segmentation methods are proposed in semiautomatic or fully automatic manner. By employing the intersection consistency in 3D space and projection consistency in 2D images, Zeng and Quan [48] proposed a silhouette extraction algorithm from multiple images of unknown background, and a silhouette carving algorithm for robust visual hull reconstruction as extension. In [30], provided the minimum user input to hardly constraint the "target object" and "background" pixels in only one of MVI, tentative segmentation of all views is achieved using traditional graph cut technique. Then, the visual hull of an object with calibrated cameras is reconstructed from silhouette of MVI, and final results are acquired by back-projection 3D model to 2D images to eliminate the segmentation errors in the tentative stage. Lee et al. [24] proposed an automatic foreground extraction method which can simultaneously identify region of interest in MVIs without any a *priori* knowledge on the background and user interaction. Driven by the initial segmentation from the intersection of viewing volumes, iterative optimal segmentation of all views is conducted using graph cut method, where the prior term in the energy function encodes the spatial consistency exploiting the multiview silhouette coherence.

5.2.3 Proposed Multiview Image Segmentation Algorithm

Detection and Localization of OOIs is the first but important step for the video tracking. To initialize the tracked OOIs in the sequence, we propose a depth-based object segmentation algorithm to automatically separate the OOIs from background and identify individual object in the initial frame of a multiview video. The segmentation algorithm is developed for different two scenarios that are the spatially separated objects and overlapped objects.

5.2.3.1 Automatic Object Extraction

The semiautomatic object extraction algorithms, which require user-supplied priors such as brush stokes [2, 8] and bounding box [25, 37] are not preferable in the MVI/V due to large quantities of data with users' interventions, thus fully automatic algorithm is in the highly demand. Automatic object extraction is still a challenging problem, especially when no prior information (background image) is provided or no semantic cues are extracted from the scene.

In our recent work [47], to automatically extract the OOIs patches for initialization of the segmentation process, saliency model is employed to compute a saliency map for the key view (middle view in 5 views) of initial frame, where higher-level features that are the depth and motion estimated off-line are utilized. The selection of these two features is based on the following reasons: human attentions are generally more focused on the moving object than the static one in the video; an OOI appears to have similar depth values in the 3D scene thus form a uniform distribution in the depth field. By thresholding, morphological operations and connected component analysis on the saliency map, initial OOIs can be automatically extracted to trigger the subsequent segmentation process. Figure 5.3 shows the saliency maps and initial OOIs in two key view images, which are used to model the foreground and background distributions.

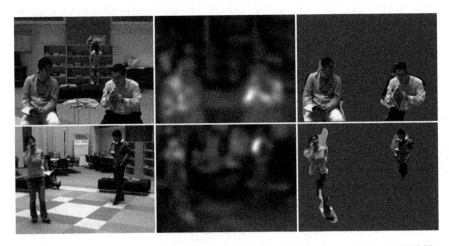

Fig. 5.3 Automatic object extraction: *left*: input image; *middle*: saliency map; *right*: initial OOIs; *top*: Reading sequence; *bottom*: Calling sequence

5.2.3.2 Segmentation of Multiple Separated Objects

In the computer vision, image segmentation can be formulated as energy minimization problem. The general formulation of energy function is given in (5.1):

$$E(f) = \sum_{(p \in P)} E_p(f_p) + \lambda \sum_{(p,q \in N)} E_{p,q}(f_p, f_q), \tag{5.1}$$

where f is the labeling field, P is the set of pixels, and N is the neighborhood system. Data term $E_p(f_p)$ is the likelihood energy and smoothness term $E_{p,q}(f_p, f_q)$ is the prior energy. λ is a parameter to balance these two terms.

Basic Energy Function for Key View Segmentation

Traditional graph-cut based segmentation using only color/contrast cues is error-prone especially on the regions with similar foreground/background features, leading to inaccurate results. It suggests a robust hybrid approach with more features.

Data term: $E_p(f_p)$ in (5.1) combines color and depth features to evaluate the likelihood of a certain pixel p in the key view images assigned to the label f_p:

$$E_p(f_p) = E_{pc}(\theta_c; z_p; f_p) + E_{pd}(\theta_d; z_p; f_p)$$
$$E_{pc}(\theta_c; z_p; f_p) = -\log g(z_p | f_p, k_p) - \log w(f_p, k_p)$$
$$E_{pd}(\theta_d; z_p; f_p) = -\log h(z_p, f_p) \tag{5.2}$$

where θ_c, θ_d are the color and depth distributions modeled by the Gaussian Mixture Model (GMM) and the histogram model $h(\cdot)$ respectively. $g(\cdot)$ denotes a Gaussian probability distribution and $w(\cdot)$ is the mixture weighting coefficient. k_p is the GMM component variable, set as 5 for foreground objects and 10 for the background. $z_p = \{d, r, g, b\}$ is a 4-dimensional feature vector for pixel p, representing the depth and the three color components.

Smoothness term: $E_{p,q}(f_p, f_q)$ in (5.1) measures the penalty of two neighboring pixels p and q with different labels and is defined as follow:

$$E_{p,q}(f_p, f_q) = dist(p, q)^{-1} \cdot \exp\{-diff(c_p, c_q)\}$$
$$diff(c_p, c_q) = \frac{1}{3}\left(\beta_r \cdot (r_p - r_q)^2 + \beta_g \cdot (g_p - g_q)^2 + \beta_b \cdot (b_p - b_q)^2\right) \tag{5.3}$$

where $dist(p, q)$ and $diff(c_p, c_q)$ are the coordinate distance and the average RGB color difference between p and q respectively. $\beta_r = (2 < \|r_p - r_q\|^2 >)^{-1}$, where $< \cdot >$ is the expectation operator for the red channel. β_g and β_b are defined similarly for the green and blue channels respectively.

Multiple Objects Segmentation Using Graph Cut

Comparing with the single object segmentation, multiple objects segmentation as a general case is investigated in our work. Based on the assumption that each object is not overlapped, we convert multiple objects segmentation into several sub-segmentation problems. For individual object, we construct a sub-graph for the pixels belonging to its "Object Rectangle", which is an enlarged rectangle of bounding box to encompass the whole object and restricts the segmentation region. Bi-label graph cut is employed to minimize the energy function and segments each object. Experimental results using basic energy function with different and combined features are shown in Fig. 5.4a–c.

Modified Energy Function for Key View Segmentation

The segmentation quality using combined features in Fig. 5.4c has outperformed the ones using either single feature in Fig. 5.4a,b. However, when the scenes contain complex background, notable segmentation inaccuracy around the objects still exists and leads to unsatisfactory results. These errors can be classified into two groups, which are highlighted with rectangle and ellipse respectively, as shown in Fig. 5.4c. To tackle these two problems, we propose a modified energy function containing two novelties: *background penalty with occlusion reasoning* is to handle the rectangle errors by refining data term in (5.2), and *foreground contrast enhancement* is to remove ellipse errors by refining smoothness term in (5.3). Based on the important observations that the focused object commonly appears in all the cameras, and background regions around the object boundary are occluded by observing the same scene from different perspective as shown in combined occlusion map CO_t^v

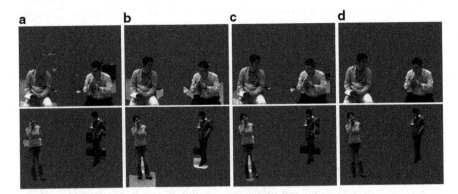

Fig. 5.4 Segmentation results using basic energy function and refinement using modified energy function: Basic energy function using (**a**) color; (**b**) depth; (**c**) combined color and depth; (**d**) results using modified energy function; *top*: Reading sequence; *bottom*: Calling sequence

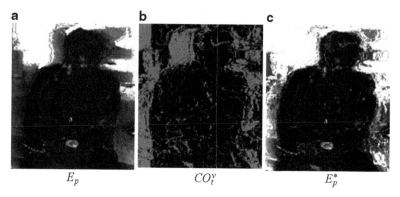

Fig. 5.5 Visualization of background penalty with occlusion reasoning

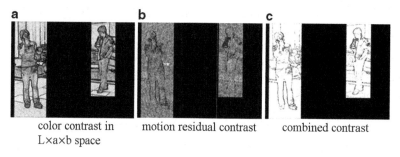

color contrast in   motion residual contrast   combined contrast
L×a×b space

Fig. 5.6 Visualization of the smoothness term in "object rectangle"

in Fig. 5.5b, background penalty with occlusion reasoning introduces a background penalty factor $\alpha_{bp} = 3.5$ to enforce the background energy of the occluded pixels:

$$E_p^*(f_p) = \alpha_{bp} \cdot E_p(f_p) \ \ (f_p = 0 \ \& \ CO_t^v(p) = 128), \tag{5.4}$$

where $f_p = 0$ and $CO_t^v(p) = 128$ if p is defined as the occluded background. E is the basic background energy and E^* is the background energy with occlusion penalty.

The motivation of foreground contrast enhancement is to enhance the contrast across foreground/background boundary and attenuate the background contrast shown in Fig. 5.6c by combining the color contrast in L×a×b space in Fig. 5.6a and motion residual contrast Fig. 5.6b. The improved results using modified energy function is illustrated in Fig. 5.4d.

Multiview Segmentation

In the above work, we have dealt with the segmentation in a single *key view* of the *initial frame*. Accurate object segmentation for all views of a frame should be provided for the further applications.

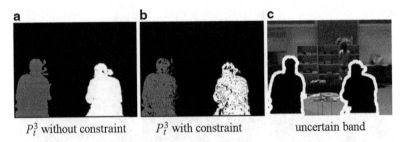

P_t^3 without constraint P_t^3 with constraint   uncertain band

Fig. 5.7 Disparity projection and uncertain band

Disparity Projection Under Visibility Constraint

Based on the segmentation result of the *key view*, the coarse predictions of the other views can be projected by pixel-based disparity compensation, which exploits the spatial consistency among interview images. However, disparity vectors cannot be estimated correctly for the occluded areas, introducing serious prediction errors as in Fig. 5.7a and the undesired effect for the subsequent process. Since only the OOIs should be projected in the target view, which are defined as visible ($CO_t^{v_i}(p) = 0$), thus the projection is performed under visibility constraint:

$$P_t^{v_i}(p) = f_t^{v_j}\left(p + D_t^{v_i,v_j}(p)\right) \quad (CO_t^{v_i}(p) = 0). \tag{5.5}$$

Uncertain Boundary Band Validation

Because of the existence of noise and non-homogeneity in the estimated field, and despite performing post-processing after the predictions, inaccuracy still exist along the object boundary. To improve the segmentation results, we construct an uncertain band along the object boundary as in Fig. 5.7c based on an *activity* measure. We define the *activity* of a pixel as the motion variance within its second-order neighborhood. The pixel with the highest *activity* is searched within the neighborhood of each contour pixel, and a band centered at the most active pixel is defined as uncertain region. The pixels lying in the inner band are labeled as foreground ($f_p = n$), and outer band pixels are background ($f_p = 0$). The indices of pixels in the uncertain band are set to be $255 - n$. Labeling field for the uncertain band is validated using graph cut to yield more accurate segmentation layers. The segmentation results of multiview images are presented in Fig. 5.8.

5.2.3.3 Segmentation of Multiple Overlapped Objects

Segmenting multiple simultaneous objects under occlusion is more difficult task than when the targets are spatially separated without overlapping regions. In this

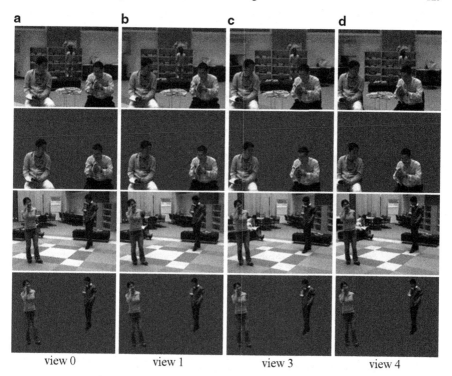

a	b	c	d

view 0	view 1	view 3	view 4

Fig. 5.8 Multiview segmentation of initial frame: *First & Second row*: Reading sequence; *Third & Fourth row*: Calling sequence; *top*: input image; *bottom*: superimposed mask

section, we discuss the proposed algorithm that first extract the foreground as multiple overlapped human objects, and then segment them into individual objects.

Adaptive Background Penalty with Occlusion Reasoning

When the segmentation starts from the initial frame with overlapped objects, not all parts of the objects in the target view are also visible in other reference views, as shown in Fig. 5.9a–c. The interview occlusions displayed in Fig. 5.9d exist not only in the transition between the object and the background (interobject occlusion), but also in the interior of the object (intraobject occlusion). In order to distinguish the interobject occlusion from intraobject occlusion, we can assign more background energy on the interobject occlusion rather than on the intraobject occlusion, adaptive penalization of the occlusion region is adopted that the penalty factor α_{bp} in (5.4) is assigned according to the motion property using the following equations:

$$\alpha_{bp}(p) = \frac{P_{\text{motion}}(m_p|f_p = 0)}{P_{\text{motion}}(m_p|f_p = 1) + \eta} \quad P_{\text{motion}}(m_p|f_p) = \log h(m_p|f_p), \qquad (5.6)$$

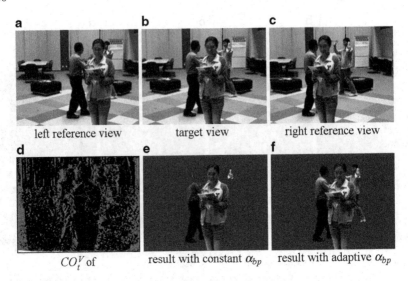

left reference view   target view   right reference view

CO_t^V of   result with constant α_{bp} result with adaptive α_{bp}

Fig. 5.9 Adaptive background penalty with occlusion reasoning

where $f_p = 0$ is for the static background and $f_p = 1$ for the moving object. m_p is the motion vector of p and $P_{\text{motion}}(m_p|f_p)$ is the motion log-likelihood of the pixel associated with the label. η is a small value to avoid the division by zero. Equation (5.6) indicates that the interview occlusion with small motion is more likely to be the interobject occlusion resulting in large value of α_{bp}, and vice versa. Figure 5.9e illustrates the segmentation result with constant α_{bp}, where the interview occlusions are equally penalized to be the background using the same factor, whereas the improved result using adaptive α_{bp} is evident in Fig. 5.9f where the background penalty is changed according to the value of α_{bp}.

Depth-Assisted Object Segmentation

Given the extracted foreground regions, object segmentation is equivalent to a k-class pixel labeling problem. By assuming that the human objects stand in the different depth layers, a coarse labeling field as shown in Fig. 5.10b can be obtained by k-means clustering of the depth map, where the number of human hypotheses is automatically determined as the number of continuous bins of the depth histogram. Due to the outliers in the estimated disparity field and the resultant reconstructed depth map, misclassifications exist in the coarse labeling field especially in the area of intraobject occlusion. We improve the initial labeling using the *depth ordering* method that If we know the layer L_1 is behind layer L_2, the occlusion region must belong to the L_1.

Because of the multiple overlapped human objects, segmentation with occlusion cannot be solved using bi-label graph cut. Given the initial results as shown

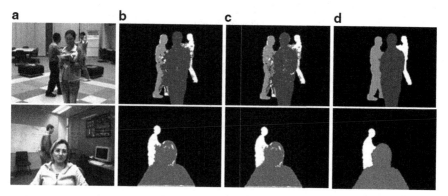

Fig. 5.10 Depth-assisted object segmentation: (**a**) initial frame, (**b**) initial labeling by depth clustering, (**c**) improved classification using depth ordering, (**d**) objects segmentation results, *top*: Three-People sequence, *bottom*: IU sequence

in Fig. 5.10c, graph cut with α-expansion is employed to minimize the energy function by fusing the color and motion cues, and segment the multiple human objects simultaneously.

5.3 Multiview Video Tracking

In the video sequence, observers generally focus more on the static or moving object than the background, and the video-based applications are commonly developed based on the object-level, thus we mainly address the object-based multiview video tracking. Comparing with the image segmentation, segmenting the sequence into spatiotemporally consistent volume and tracking meaningful video object across frames increase the complexity of video segmentation and tracking. Segmenting and tracking multiple moving objects automatically and accurately in wide-range environment or with largely crowded individuals is more challenging task. Using monocular camera is insufficient due to the limited field of views and significant occlusion. stereo/multiple cameras are reasonable alternatives to solve this difficult problem by data association across-time and across-view. The current works on multiview video segmentation and tracking are categorized into two main classes, namely feature-based and homogrophy-based, where segmentation of video objects are described as different representations such as bounding box, 2D/3D ellipse, or foreground mask. The categorical overview of multiview video segmentation and tracking approaches is shown in Fig. 5.11. Feature-based approaches perform segmentation and tracking by fusing various 2D/3D features and/or introducing statistic filters. Homogrophy-based approaches incorporate homography constraint across multiple views into the segmentation and tracking framework.

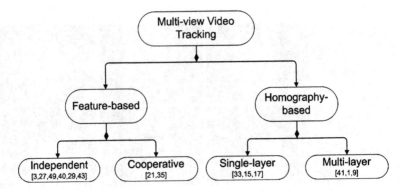

Fig. 5.11 Overview of multiview video tracking approaches

Table 5.1 Feature-based tracking approaches

		Ref.	Features	Tracker	Rep.
IT	NF	[3]	Color, depth, motion	No	FM
	FCV	[27]	Color, height, width, stereo	Kalman filter	BB
		[49]	Shape, appearance, depth, motion	Kalman filter	BB
		[40]	Appearance, depth	Bayer tracker	BB
	FCT	[29]	Luminance, color, human template, 2D position	Kalman filter	BB
		[43]	Stereo, 2D position, intensity	Kalman filter	FM
CT		[21]	Color histogram, 4D entities of rectangle	Particle filter	BB
		[35]	5D state space of ellipse	Particle filter	EM

IT: Independent tracking; *CT*: Cooperative tracking; *NF*: No fusion; *FCV*: Fusion to common view; *FCT*: Fusion to common tracker; *Ref.*: Reference; *Rep.*: Representation; *FM*: Foreground mask; *BB*: Bounding box; *EM*: Ellipse model

5.3.1 Feature-Based Tracking

Feature-based tracking methods employs feature match framework with two steps: feature extraction and feature matching. Feature-based tracking can be performed either independently in each view or cooperatively across views. A classified summary of feature-based tracking approaches is presented in Table. 5.1.

5.3.1.1 Independent Tracking

Independent tracking firstly implements single camera detection, segmentation, and tracking on its own view. Based on whether the multi-camera fusion module is involved or not, the independent tracking is further divided into *independent tracking without fusion* and *independent tracking with fusion*.

Independent Tracking Without Fusion

Independent tracking without fusion is to track the regions of object in multiview video by segmentation of each frame in individual view.

Cigla et al. [3] presented a multiview video object segmentation algorithm by integrating color, depth, and motion features. A region-based color segmentation algorithm based on modified Normalized Cuts is firstly adopted to generate over-segmented segments. Depth map is then estimated for subregions in the available segmentation mask by region-wise planarity assumption. Multiview video segmentation is extended from image segmentation by combining the color and depth with additional optical flow information to provide the motion field.

Independent Tracking with Fusion

Independent tracking with fusion is to segment the tracks in each camera stream and then project the tracks to another camera view or a common view (ground plane [27, 49], "plan-view" [40]), or collect the 2D local tracks from individual view to a global 3D track [29] or central node [43].

A multiview segmentation and tracking system in cluttered scene with multiple people is presented in [27], which is named M2Tracker. Exploiting the approximate object's shape and location prior helps the segmentation of each view using Bayesian classifications. The region-based stereo algorithm is capable of finding the 3D points inside the object. By combing evidences from different camera pair and producing feet-region likelihood estimation on the ground plane, globally optimum detection and tracking of object is attainable using Kalman filter. Zhao et al. [49] presented a similar and realtime system that detects and tracks object independently for each stereo camera, and integrate tracking results from all camera pairs to a multi-camera tracker (McTracker), which track each object on the ground plan. An object tracking framework based on dynamic Bayesian formulation is reported in [40] to observe and track object on the plan-view map by combining local appearance feature and stereo depth data.

Instead of projecting the multiple single-view tracks to a common view, combining the tracked 2D object into a 3D tracking module is another strategy for multiview data fusion. In [29], following the people detection using background subtraction and human-template correlation, 2D objects are tracked separately in each of camera by a graph matching. A 3D tracker is established using geometrical consistency between 2D objects to estimate the 3D head position. For tracking large numbers of tightly-spaced and rapid-moving objects, i.e., hundreds of flying bats, a multiobject multi-camera tracking framework is proposed in [43]. It maintains the sensor-level tracking in each view and single-view measurements send to a central node for across-view data association and tracker fusion. The feedback from central node is then used for adjusting sensor track with across-frame data association.

5.3.1.2 Cooperative Tracking

In the cooperative mechanism for multiple camera tracking, the individual target is tracked not only by the measurement in its own camera view, but also through the camera interactions from its counterpart in other cameras.

Occlusion increases the complexity in tracking multiple targets. Since single viewpoint loses the depth and occlusion information, multiple cooperative/collaborative cameras are indeed helpful to maintain the tracking performance since the tracking process of target in the visible view can assist the process in the occluded view. Lien and Huang [21] proposed a multiview based cooperative tracking of multiple human objects. Each object in each view is tracked using particle filter, and occlusion information is revealed by homography transformation in different views. The cooperative tracking model integrates the tracking results across views and applies human interaction between different views, which is proved to be more effective than non-cooperative tracking system. Qu et al. [35] presented a Bayesian framework for multiple-target tracking using multiple collaborative cameras. *Camera collaboration* integrated into a graphical model links the target's state in the analyzed camera view and its counterpart state in another view, and is activated by "need-driven"-based scheme to handle multitarget occlusion as it happens.

5.3.2 Homogrophy-Based Tracking

Locating and tracking dense object in crowded scene is a challenging problem due to the significant occlusion and extensive motion. Using planar homography constraint [14] to collect multiview information is a relatively new area, which helps improve the segmentation and tracking performance towards high accuracy and robustness in such difficult situation.

5.3.2.1 Single-Layer Homography

Applying single-layer homography is to project the individual track into a visual plane through homographical transform, which is generally the ground plane from the top down view in the scene, to detect and track the feature points of object on the ground (Fig. 5.12).

Park and Trivedi [33] presented a homography-based tracking framework for analysis of people and vehicle activities in the crowded scene. Multiple views of the same object are projected onto a common planar homography map to detect the footage regions of objects. The detected objects are tracked in the homography domain using Kalman filter to estimate the position, size, and velocity of persons and vehicles. In [15], foreground blobs are segmented using human appearance from background subtraction. To precisely locate the ground locations of the multiple people possibly experiencing occlusion, center vertical axes of the person across

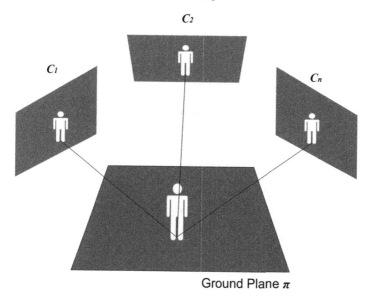

Fig. 5.12 Detection and tracking using single-layer homography: projection from multiple views to the ground plane of top-down view

views are homographically mapped into the top view ground plane and the intersections are estimated as the feet point. Multi-hypothesis trackers are established using particle filters for robust tracking. A similar work is reported by Khan and Shan [17], where tracking multiple people is solved by graph cut segmentation of spatiotemporal feet blob volume.

5.3.2.2 Multi-Layer Homography

To enhance the accuracy of localization and reduce the false alarm, it can exploit multiplanar homography constraint to combine projections in multiple layers that are parallel to the ground plane (Fig. 5.13).

Tong et al. [41] proposed a multicamera approach for multipeople localization using multiplanar homography constraint. Foreground regions are segmented using Gaussian mixture model-based background suppression for each view and each frame, which are warped to the reference view to get target section on the plane. Five-planar homography from ground plane to head plane is adopted to gather all plane information to final overlooking view, where the people are clustered for localization. A homography framework in [1] is developed for multicamera tracking. Foreground blobs are extracted using graph-cut based foreground sub-traction. Applying multi-layer homography results in the increasing reliability of localization by transforming the foreground map of reference view with multiple layers, where the feet positions are detected to indicate the coherent foreground regions. Instead of tracking footage of people, Eshel and Moses [9] worked on tracking people's head

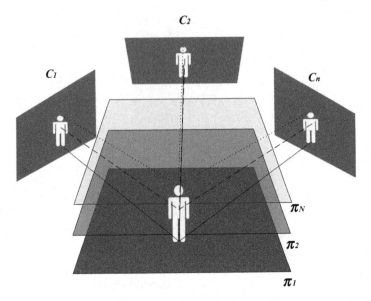

Fig. 5.13 Detection and tracking using multi-layer homography: projection from multiple views to several parallel planes

using high elevated cameras. Multi-height homography of three planes parallel to the floor is involved to detect the head top of the people, which is the feature point to track the human motion.

5.3.3 Proposed Multiview Video Tracking Algorithm

Unlike the approaches only tracking the interest points (feet, head) of the target, we propose a feature-based tracking method that define the entire object region as the tracker and update the deformable objects' regions frame by frame with various dynamic motions.

5.3.3.1 Tracking of Multiple Separated Objects

Segmenting the consecutive frame is achievable as the motion information is known. Motion prediction is a form of tracking, which enforces the temporal consistency between adjacent frames in video. The coarse prediction of the current frame is projected by pixel-based motion compensation from the mask of its previous frame. Uncertain band construction and validation for individual object as in Sect. 5.2.3.2 results in the projection refinement. The excellent performance of the tracking strategy has been demonstrated on a couple of real and complex videos containing spatially separated objects as illustrated in Fig. 5.14.

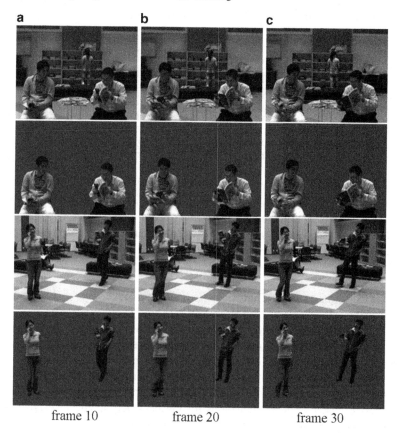

Fig. 5.14 Video tracking of separated objects: *First & Second row*: Reading sequence; *Third & Fourth row*: Calling sequence; *top*: input frame; *bottom*: superimposed mask

5.3.3.2   Tracking of Multiple Overlapped Objects

Accurate and consistent tracking multiple overlapped objects is more difficult problem than that of spatially separated objects resulting from the dynamic change of objects attributes such as appearance, shape, and visibility.

Tracking Using Basic Strategy

Due to the dynamic movements of the human objects, the motion field between adjacent frames cannot be estimated accurately, especially between the intersections of different object layers caused by the motion occlusion without correspondence. These motion compensation errors degrade the segmentation results, which are accumulated into the following frames. Figure 5.15b shows the tracking results after

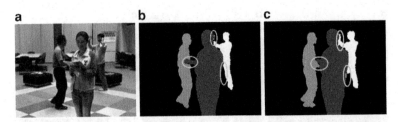

Fig. 5.15 Objects tracking after six frames: (**a**) input image, (**b**) mask by only motion compensation and uncertainty analysis, (**c**) improved result by motion occlusion with layer transition

six frames of initial frame using only basic strategy in Sect. 5.3.3.1. In the object mask shown in Fig. 5.15b, the new uncovered regions across different object layers are lost or mislabeled, resulting from the prediction and segmentation errors accumulating frame after frame.

Motion Occlusion As Layer Transition

Tracking the focused candidate regions by only motion compensation and uncertainty analysis with objects' overlapping introduces errors because of the motion occlusion even in the newly exposed regions. To handle this problem, we model the motion occlusion as layer transition, since the emergence of occlusion is always accompanied by label transition between different object layers. We now discuss two distinct classes of layer transitions for the occluded pixels corresponding to *background to be covered* and *uncovered new regions*.

Background to Be Covered

If the pixel in the previous frame is labeled as background layer ($f_p^{t-1} = 0$), it will only transit to a certain foreground object in the current frame.

The determination of the object index is formulated as a Bayesian maximum a posteriori (MAP) problem:

$$f_p^t = \arg \max_{f_p \in F_{\text{foreground}} = \{1,2,\ldots,N\}} P(f_p|x_p), \qquad (5.7)$$

where f_p^t is the label of pixel p in the current frame at time instance t. $F_{\text{foreground}}$ is the foreground label set and N is the number of objects. According to the Bayesian rule, the posterior probability $P(f_p|x_p)$ that an observation of pixel x_p belonging to an object can be decomposed into a joint likelihood function $P(x_p|f_p)$ and a prior $P(f_p)$ is given as:

$$P(f_p|x_p) \propto P(x_p|f_p)P(f_p). \qquad (5.8)$$

By assuming the uniform distribution of prior $P(f_p)$, the MAP problem is reduced to a maximum likelihood (ML) problem to maximize the joint likelihood function $P(x_p|f_p)$, which is evaluated using the color cue modeled by the GMM, combined with the depth and motion cues modeled using the histogram:

$$P(x_p|f_p) = P_{color}(c_p|f_p) + P_{depth}(d_p|f_p) + P_{motion}(m_p|f_p)$$

$$= \log g(c_p|f_p) + \log h(d_p|f_p) + \log h(m_p|f_p), \qquad (5.9)$$

where c_p is RGB color channels of p, d_p, and m_p denote the depth and motion features of p respectively.

Uncovered New Regions

If the pixel in the previous frame is labeled as foreground object ($f_p^{t-1} \in F_{fore}$), it will only transit to the intersected same layer or the back layer.

Similarly, finding the corresponding layer is an ML problem:

$$f_p^t = \arg \max_{f_p \in F_{feasible} = \{(f_p^t \geq f_p^{t-1} \cup f_p^t = 0) \cap Ins(f_p^{t-1})\}} P(x_p|f_p), \qquad (5.10)$$

where f_p^{t-1} is the label of p in the previous frame at time $t-1$. $Ins(f_p^{t-1})$ is defined as the set of layer that f_p^{t-1} intersects with, and $F_{feasible}$ is the feasible label set for a foreground object, which is located in the same or back layer in $ins(f_p^{t-1})$. The transition between the same layer corresponds to the new uncovered part of the object. The transition from the front layer to the back layer indicates the exposure of occluded part.

Feature selection: For the uncovered regions appearing on the scene, the new exposed parts may not be consistent with its associated object, eventhough they are very similar to the other object. Under such condition, the color component in the joint likelihood function in (5.9) will mislead the label decision, which makes the color evidence invalid. To avoid this from happening, we select the appropriate features in the evaluation of joint likelihood function based on the statement of new uncovered regions described in (5.10). Since the label that will be transited to should exist in $F_{feasible}$, we traverse all the possible labels to measure the color likelihood, and find the label that corresponds to the maximum color likelihood which is not in the feasible set $F_{feasible}$:

$$f_p^t = \arg \max_{f_p \in F_{foreground \cup 0}} P_{color}(c_p|f_p), f_p^t \notin F_{feasible}. \qquad (5.11)$$

This bias indicates that new uncovered parts have nonhomogenous appearance with the associated object. Thus, we remove the color term and retain the depth and motion terms in (5.9). Otherwise, we combine color and depth cues to calculate $P(x_p|f_p)$, which is distinctive enough to make a good decision.

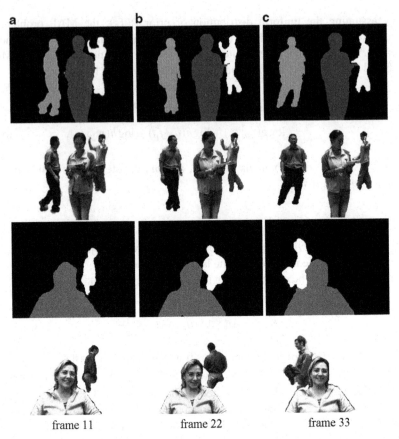

frame 11   frame 22   frame 33

Fig. 5.16 Video tracking of overlapping objects: *First & Second row*: Three people sequence; *Third & Fourth row*: IU sequence; *top*: mask; *bottom*: superimposed mask

From the comparison of Fig. 5.15b,c, the tracking errors in Fig. 5.15b without motion occlusion analysis have been successfully handled using the proposed tracking strategy towards modeling the motion occlusion as layer transition, which can achieve more precise representation of individual object by eliminating the lost regions or mislabeling of the new uncovered parts. Object tracking results on every 11th frame on *Three-People* and *IU* sequences are shown in Fig. 5.16. The selected frames show the average performance of video tracking, which contain typical tracking problems such as objects' partial occlusion, separation, and appearance of new part. 2D regions of each object are tracked consistently and correctly across the video sequences with various dynamics in the scene. The satisfactory segmentation and tracking results presented in Fig. 5.16 demonstrate the efficiency and robustness of the proposed algorithm in subjective performance.

5.4   Conclusions

In this chapter, we reviewed and categorized the representative and state-of-the-art approaches for multiview image segmentation and video tracking. The first part of this chapter dedicates on two classes of multiview image segmentation techniques, namely region-based and object-based segmentation. We pay more attention to the latter approach due to the popularity of accessing and manipulating scene on the object level, which can be further grouped into depth-based and silhouette-based approaches. In the second part, we discuss two classifications of multiview video tracking methods that are the feature-based and homography-based tracking. Feature-based tracking focus on integrating various 2D/3D features and statistic filters into the tracking framework, while the homography-based algorithms perform detection and tracking by exploiting single-layer or multi-layer homography constraint to achieve multiview data fusion. Additionally, our proposed multiview object-based segmentation and feature-based tracking algorithms are presented in the corresponding sections, following the experimental results to demonstrate the excellent performance with high accuracy and robustness.

References

1. Arsic D, Lyutskanov A, Rigoll G, Kwolek B (2009) Multi camera person tracking applying a graph-cut based foreground segmentation in a homography framework. IEEE International Workshop on Performance Evaluation of Tracking and Surveillance
2. Boykov Y, Jolly MP (2001) Interactive graph cuts for optimal boundary and region segmentation of objects in N-D images. IEEE International Conference on Computer Vision: 105-112
3. Cigla C, Aydin Alatan A (2008) Depth Assisted Object Segmentation in Multi-View Video. 3DTV Conference: The True Vision – Capture, Transmission and Display of 3D Video: 185-188
4. Chai D, Ngan KN (1999) Face segmentation using skin color map in videophone applications. IEEE Transactions on Circuits and Systems for Video Technology 9(4): 551-564
5. Cheung GKM, Baker S, Simon C, Kanade T (2003) Visual hull alignment and refinement across time: A 3D reconstruction algorithm combining shape-from-silhouette with stereo. IEEE Conference on Computer Vision and Pattern Recognition 2: 375-382
6. Campbell NDF, Vogiatzis G, Hernndez C, Cipolla R (2008) Automatic 3D object segmentation in multiple views using volumetric graph-cuts. Image and Vision Computing 28(1):14-25
7. Cui C, Zhang Q, Ngan KN (2009) Multi-view Video Based Object Segmentation - A Tutorial. ECTI Transactions on Electrical Engineering, Electronics and Communications 7(2):90-105
8. Duchenne O, Audibert JY, Keriven R, Ponce J, Segonne F (2008) Segmentation by transduction. IEEE Conference on Computer Vision and Pattern Recognition: 23-28
9. Eshel R, Moses Y (2008) Homography based multiple camera detection and tracking of people in a dense crowd. IEEE Conference on Computer Vision and Pattern Recognition
10. Goldlucke B, Magnor MA (2003) Joint 3D-Reconstruction and Background Removal Separation in Multiple Views using Graph Cuts. IEEE Conference on Computer Vision and Pattern Recognition: 683-688
11. Grammalidis N, Bleris L, Strintzis MG (2002) Using the expectation-maximization algorithm for depth estimation and segmentation of multi-view images. International Symposium on 3D Data Processing Visualization and Transmission: 686-689

12. Grauman K, Shakhnarovich G, Darrell T (2003) A Bayesian Approach to Image-Based Visual Hull Reconstruction. IEEE Conference on Computer Vision and Pattern Recognition

13. Guillemaut JY, Kilner J, Hilton A (2009) Robust Graph-Cut Scene Segmentation and Reconstruction for Free-Viewpoint Video of Complex Dynamic Scenes. IEEE International Conference on Computer Vision

14. Hartley R, Zisserman A (2000) Multiple View Geometry in Computer Vision. Cambridge University Press New York, NY, USA

15. Kim K, Davis LS (2006) Multi-Camera Tracking and Segmentation of Occluded People on Ground Plane using Search-Guided Particle Filtering. European Conference on Computer Vision: 98-109

16. Kim C, Hwang JN (2002) Object-based video abstraction for video surveillance systems. IEEE Transactions on Circuits and Systems for Video Technology 12: 1128-1138

17. Khan SM, Shah Mubarak (2009) Tracking Multiple Occluding People by Localizing on Multiple Scene Planes. IEEE Transactions on Pattern Analysis and Machine Intelligence 31(3): 505-519

18. Kolmogorov V, Zabih R (2002) Multi-camera Scene Reconstruction via Graph Cuts. European Conference on Computer Vision: 82-96

19. Kolmogorov V, Criminisi A, Blake A, Cross G, Rother C (2006) Probabilistic fusion of stereo with color and contrast for bi-layer segmentation. IEEE Transactions on Pattern Analysis and Machine Intelligence 28(9): 1480-1492

20. Laurentini A (1994) The visual hull concept for silhouette-based image understanding. IEEE Transactions on Pattern Analysis and Machine Intelligence 16(2): 150-162

21. Lien KC, Huang CL (2008) Multiview-Based Cooperative Tracking of Multiple Human Objects, EURASIP Journal on Image and Video Processing, Article ID 253039

22. Lin HY, Wu JR (2008) 3D reconstruction by combining shape from silhouette with stereo. International Conference on Pattern Recognition: 1-4

23. Li Y, Sun J, Shum HY (2005) Video object cut and paste, ACM Transactions on Graphics 24: 595-600

24. Lee W, Woo W, Boyer E (2007) Identifying Foreground from Multiple Images. IEEE Asian Conference on Computer Vision: 1-10

25. Lempitsky V, Kohli P, Rother C, Sharp T (2009) Image Segmentation with A Bounding Box Prior. IEEE International Conference on Computer Vision: 277-284

26. Medeiros MAA, Cruz LADS (2008) Iterative disparity estimation and image segmentation. Conference on Signal processing, computational geometry and artificial vision: 67-72

27. Mittal A, Davis LS (2003) M2Tracker: A Multi-View Approach to Segmenting and Tracking People in a Cluttered Scene. International Journal of Computer Vision 51(3): 189-203

28. Mori G, Ren X, Efros AA., Malik J (2004) Recovering Human Body Configurations: Combining Segmentation and Recognition. IEEE Conference on Computer Vision and Pattern Recognition: 326-333

29. Mohedano R, Del-Bianco CR, Jaureguizar F, Salgado L, Garcia N (2008) Robust 3D people tracking and positioning system in a semi-overlapped multi-camera environment. IEEE International Conference on Image Processing: 2656-2659

30. Min SK, Kim JW, Park AJ, Hong GJ, Jung KC (2008) Graph-Cut Based Background Subtraction Using Visual Hull in Multiveiw Images. Digital Image Computing: Techniques and Applications: 372-377

31. Min DB, Kim DH, Yun SU, Sohn KH (2009) 2D/3D freeview video generation for 3DTV system. Signal Processing: Image Communication 24(1-2): 31-48

32. Nguyen HT, Do MN (2005) Image-based rendering with depth information using the propagation algorithm. IEEE International Conference on Acoustics, Speech, and Signal Processing

33. Park S, Trivedi MM (2007) Homography-based Analysis of People and Vehicle Activities in Crowded Scenes. IEEE Workshop on Applications of Computer Vision

34. Pylvanainen T, Roimela K, Vedantham R, Itaranta J , Wang R , Grzeszczuk R (2010) Automatic Alignment and Multi-View Segmentation of Street View Data using 3D Shape Priors. International Symposium on 3D Data Processing Visualization and Transmission

35. Qu W, Schonfeld D, Mohamed M (2007) Distributed Bayesian Multiple Target Tracking in Crowded Environments Using Multiple Collaborative Cameras. EURASIP Journal on Applied Signal Processing 2007(1): 21-21

36. Quan L, Wang JD, Tan P, Yuan L (2007) Image-based modeling by joint segmentation, International Journal of Computer Vision 75(1): 135-150

37. Rother C, Kolmogorov V, Blake A (2004) Grabcut: Interactive foreground extraction using iterated graph cuts. ACM Transactions on Graphs: 309-314

38. Sikora T (1997) The MPEG-4 video standard verification model. IEEE Transactions on Circuits System and Video Technology 7: 19-31

39. Tasi YP, Ko CH, Hung YP, Shih ZC (2007) Background Removal of Multiview Images by Learning Shape Priors. IEEE Transactions on Image Processing 16: 2607-2616

40. Tang F, Harville M, Tao H, Robinson IN (2008) Fusion of local appearance with stereo depth for object tracking. IEEE Conference on Computer Vision and Pattern Recognition Workshops

41. Tong X, Yang T, Xi R, Shao D, Zhang X (2009) A Novel Multi-planar Homography Constraint Algorithm for Robust Multi-people Location with Severe Occlusion. International Conference on Image and Graphics: 349-354

42. Wills J, Agarwal S, Belongie S (2003) what went where. IEEE Conference on Computer Vision and Pattern Recognition

43. Wu Z, Hristov NI, Hedrick TL, Kunz TH, Betke M (2009) Tracking a Large Number of Objects from Multiple Views. IEEE International Conference on Computer Vision

44. Xiao J, Quan L (2009) Multiple View Semantic Segmentation for Street View Images. IEEE International Conference on Computer Vision

45. Yang W, Ngan K, Lim J, Sohn K (2005) Joint Motion and Disparity Fields Estimation for Stereoscopic Video Sequences. Signal Processing: Image Communication 20(3): 265-276

46. Zitnick CL, Kang SB (2007) Stereo for Image-Based Rendering using Image Over-Segmentation, International Journal of Computer Vision 75(1): 49-65

47. Zhang Q, Ngan KN (2010) Multi-view Video Segmentation Using Graph Cut and Spatiotemporal Projections. Journal of Visual Communication and Image Representation 21(5-6): 453-461

48. Zeng G, Quan L (2004) Silhouette extraction from multiple images of an unknown background. IEEE Asian Conference on Computer Vision: 1-10

49. Zhao T, Aggarwal M, Kumar R, Sawhney H (2005) Real-time wide area multi-camera stereo tracking. IEEE International Conference on Computer Vision 2: 976-983

Chapter 6
Applications of Video Segmentation

E. Izquierdo and K. Vaiapury

Abstract Segmentation is one of the important computer vision processes that is used in many practical applications such as medical imaging, computer-guided surgery, machine vision, object recognition, surveillance, content-based browsing, augmented reality applications, *etc.*. The knowledge to ascertain plausible segmentation applications and corresponding algorithmic techniques is necessary to simplify the video representation into a more meaningful and easier form to analyze. This is because expected segmentation quality for a given application depends on the level of granularity and the requirement that is related to shape precision and temporal coherence of the objects.

6.1 Introduction

With the rapid growth of video data, management, access, and retrieval of desired information from humongous video library is becoming a headachy experience for users. Segmentation is one of the important computer vision processes that is used in many practical applications such as medical imaging, computer-guided surgery, machine vision, object recognition, surveillance, content-based browsing, augmented reality applications, *etc.* The knowledge to ascertain plausible segmentation applications and corresponding algorithmic techniques is necessary to simplify the video representation into a more meaningful and easier form to analyze. In fact, expected segmentation quality for a given application depends on the level of granularity and the requirement that is related to shape precision and temporal coherence of the objects. In this chapter, we discuss key applications of video segmentation. Video Segmentation refers to the process of splitting videos into homogenous spatial temporal segments meaningful from a semantic point of view. Video is a sequence of frames that have a high degree of temporal correlation among them [21]. Each frame is an image in 2D spatial plane. Though the underlying segmentation process is the

E. Izquierdo (✉)
Department of Electronic Engineering, Queen Mary, University of London, London, UK
e-mail: ebroul.izquierdo@elec.qmul.ac.uk

K.N. Ngan and H. Li (eds.), *Video Segmentation and Its Applications*,
DOI 10.1007/978-1-4419-9482-0_6, © Springer Science+Business Media, LLC 2011

Table 6.1 Segmentation
quality matrix *[1-highly
desirable, 4-least desirable]*

Scenario	Real time	Offline
User Interactive	2	4
Non-User Interactive	1	3

same, the extra time dimension in video makes segmentation in video different from that in images. As stated by Zivkovic et al. [23] the vital problem in video analysis is segmenting foreground object from background and in general, extracting high level semantics from video could be a task of interest.

Ideally, any video segmentation application should address and satisfy the following two different essential properties:

1. Precision of object contours: This refers to how well the object boundaries are correctly identified.
2. Temporal coherency of the partition: This refers to ability of segmentation algorithm to identify the segment throughout the time to enable tracking.

Sometimes, very precise contours and high temporal coherency is required. Other times, a rough identification of the object locations (e.g., using bounding boxes) is enough. The key issue is to have accurate contour with consistent partition along time. While segmentation quality is ultimate key factor to rate the performance of any method, time and need for user interaction are important concerns to be considered for segmentation. In [9], Izquierdo et al. has explained key components of segmentation system.

Correia et al. has classified applications of video segmentation into set of scenarios according to application constraints and goals [3]:

1. Real-time Nonuser Interactive scenario
2. Real-time User Interactive scenario
3. Offline User Interactive scenario
4. Offline Nonuser Interactive scenario

Li et al. [13] has classified video application into six categories such as Video Surveillance, Content-based Video Summarization, Content-based Coding, Computer Vision, Videoconferencing/Videophone applications, and Digital Entertainment.

As one can see from Table 6.1, in addition to preserving the accuracy, any highly desirable segmentation system should be (a) fully automatic, (b) able to work in real time. The score in Table 6.1 reflects the same.

6.2 Recent Trends in Video Segmentation

Recently, there has been a strong surge for video segmentation in mobile media applications such as augmented reality and mobile media communications such as object based coding. A full classification of segmentation applications into a set of

Table 6.2 Application and related work

Applications	Reference work
Visual surveillance and, traffic control	[10, 19]
Object based video coding and, event based scalable coding	[16, 22]
Three-D reconstruction(visual hull), video tooning and, rendering	[5, 20]
Augmented Reality, tourism, games and, surgery	[17, 24, 25, 27]
Content based Video Summarization	[2]
Video conferencing and video phoning application	[1]

scenarios, according to different application constraints and goals can be found in [3]. There have been a lot of work in video segmentation using colour, motion based methods [10, 19].

We have summarized the list of potential applications and related work in Table 6.2.

In fact, there are other applications like video classification, which deals with problem of categorizing a given videos sequence into one or predefined video genre. For example one might be interested in finding all non identical duplicate videos having some personality as a focus [21].

There is a paradigm shift from traditional segmentation to using depth [20], attention and prior model information [14] in addition to color and motion-based approaches [18].

Clearly, depth based approaches bear the potential discriminative power of ascertaining whether the object is nearer of farer. We have proposed and evaluated a GrabCut segmentation technique based on combination of colour and depth information [20].

However, GraphCut techniques demand user initialization. As stated in [4], while using GraphCut techniques, attention based models can be used instead of manual initialization for segmentation process. The attention models can be based on saliency map approaches, which leads to saliency-based segmentation model. However, modeling visual attention models is still a challenging problem. Visual attention models have been widely used in many applications.

In fact, to find what object is attended to and where the attention likely to be, filtering and prioritizing the information is vital. This is analogous to the nature of human fovea, which acts according to stimulus. There are two major computational models of attention such as:

1. *Bottom-up attention*: It is based on combination of low level features which include both oriented as well as non-oriented features such as colour, contrast, and orientation.
2. *Top-down attention*: It involves task dependent processing, which generally requires some prior knowledge about the scene. In effect, the user attention is guided by what he sees.

Itti et al. [8] has proposed a model for finding low-level surprise at every location in video streams. The method correlates with gaze shifts of two human observers watching complex video clips such as television programs.

Model based approaches for video segmentation has been used in works such as [12,15]. As stated by Hampapur et al. video segmentation requires explicit model of video [7]. In industrial applications, prior computerized base models are designed using AutoCAD or CATIA. This knowledge can be used to have model aided object segmentation. For example, Li et al. [14] has proposed face segmentation based on saliency model.

6.3 Object Based Surveillance Analysis

Surveillance allows identifying any abnormal activity in a given environment thereby enhancing public safety and reducing crime. If there are more number of smart cameras used in surveillance process, then there is a risk that person who monitors may not be able to analyse the videos effectively. As stated by Li et al. surveillance helps to anticipate and reveal patterns of their actions and interactions with one another in their environment to determine when "alerts" should be posted to security unit [13]. Hence there is a need for inspection process such as detecting unattended bags, loitering people and any suspicious activity. For example, as stated in [26], Cromatica has been tested at London Liverpool station. Cromatica is based on an algorithm, which detects differences in frames. For example, if there is too much movement in images, then abnormal behaviour event alert might be raised. In fact, it is easier to segment moving objects in video sequences unlike static objects in images. Zgaljic et al. has described surveillance centric codec for industrial applications [22]. It includes target detection recognizing a target instead of segmenting precisely refer Fig. 6.1. As it can be seen from Fig. 6.2, there exists application to track car in the jammed or congested park lot. Gelesca et al. has done a brief study on evaluation of algorithms used for surveillance [6]. Video surveillance systems concerning algorithms for tracking moving object can be of two types: bounding box and/or centre of gravity. In surveillance process, video frame is analyzed and object location is retrieved as a function of time. The flaws include over segmen-

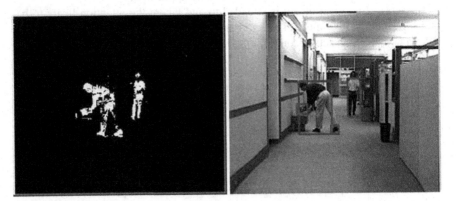

Fig. 6.1 Hall Sequence 110th frame and tracking result [22]

Fig. 6.2 Finding
Vehicles [28]

tation, under segmentation (border holes), Flickering etc. The detection and false alarms rates are estimated by counting how many times interesting and irrelevant regions are detected.

6.4 Object-Based Scalable Video Coding

Surveillance centric methods reduce video bit rate without jeopardising information relevant to surveillance application. As stated in [16], H.264/MPEG-4 AVC provides a fully scalable extension, SVC, which achieves significant compression gain and complexity reduction when scalability is sought, compared to the previous video coding standards. Krutz et al. has described a methodology to separate a video scene into shots that are coded either with an object-based codec or the common H.264/AVC [11]. Their strategy is to use different video codecs for different kinds of content in order to obtain higher coding gain. Contrasting conventional coders, the system [22] addresses the requirements of surveillance application scenarios. It aims at achieving bit-rate optimization, and adaptation of surveillance videos for storing and transmission purposes. In the system, the encoder communicates with a Video Content Analysis module that detects events of interests in videos captured by CCTV. Bit-rate optimization and adaptation is achieved by exploiting scalability properties of the target codec. Temporal segments containing events relevant to surveillance application are encoded using high spatial temporal resolution and quality while the portions irrelevant from the surveillance standpoint are encoded at low spatial temporal resolution and/or quality. The model optimises the bit allocation between a wavelet-based scalable video coder and a forward error correction codes.

Ramzan et al. [16] has classified the temporal segments of the video sequence into two types:

- Temporal segments representing an essentially static scene (e.g. only random environmental motion is present – swaying trees, flags moving on the wind, etc.)
- Temporal segments containing nonrandomised motion activity (e.g. a vehicle is moving in a forbidden area).

To enable the above classification, background subtraction and tracking module is used as Video Content Analysis (VCA). It uses a mixture of Gaussians to separate the foreground from the background. Each pixel of a sequence is matched with each weighted Gaussian of the mixture. If the pixel value is not within 2.5 standard deviations of any Gaussians representing the background, then the pixel is declared as the foreground. Since the mixture of Gaussians is adaptive and more than one Gaussians are allowed to represent the background; this module is able to deal robustly with light changes, bimodal background like swaying trees and introduction or removal of objects from the scene. The output of the module defines parameters of compressed video, which is encoded with the W-SVC framework.

Further, three different scalability issues are discussed:

1. *Temporal scalability*: It refers to the possibility of reducing the temporal resolution of encoded video directly from the compressed bit-stream, i.e. number of frames contained in one second of the video.
2. *Spatial scalability*: It refers to the possibility of reducing the spatial resolution of the encoded video directly from the compressed bit-stream, i.e. number of pixels per spatial region in a video frame.
3. *Quality scalability*: This refers to the possibility of reducing the quality of the encoded video. This is achieved by extraction and decoding of coarsely quantised pixels from the compressed bit-stream. This is also called as SNR (Signal-to-Noise-Ratio) scalability or fidelity scalability.

As shown in Fig. 6.3, if object is detected, then high resolution video is transmitted else low quality video is transmitted or no transmission is done (Figs. 6.4, 6.5, 6.8, and 6.9).

6.5 3D Reconstruction

Using silhouettes and camera calibration parameters, 3D visual hull can be reconstructed [5]. Now, we explain a depth-based segmentation technique GrabcutD to obtain visual silhouettes.

Fig. 6.3   Event based scalable coding (Event occurs) [16]

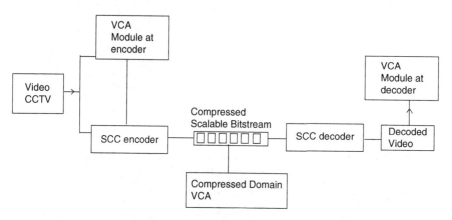

Fig. 6.4   Scalable video codec architecture [22]

6.5.1   GrabcutD: Modified GrabCut Using 4 Channels

GrabcutD is a modified version of Grabcut using depth-based information. In [20], for experimentation purposes, we used the video dataset frames from MSR ballet sequence dance video. Initially, for each frame, users select a bounding box and the pixels inside and outside rectangle are represented by foreground and background classes respectively. From each trimap selection of the foreground and background, the histograms are formed using 4 channels information (*Red, Green,*

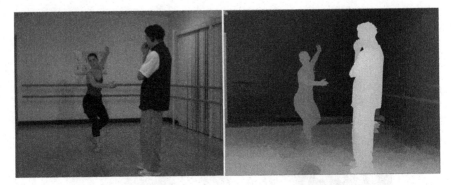

Fig. 6.5 Ballet sequence Image [29]

Fig. 6.6 Existing method
(GrabCut) results for Ballet
sequence [20]

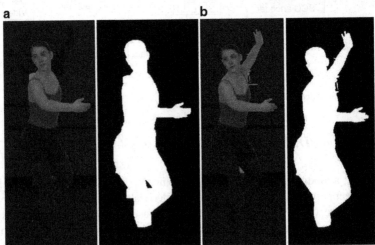

Fig. 6.7 Ballet sequence Dancer: (**a**) GrabCut and (**b**) GrabCutD (Color and Depth) [20]

Fig. 6.8 Augmented Game application [27]

Fig. 6.9 Panoramascope [24]

Blue, and Depth) instead of just colour *(Red, Green, and Blue)*. The Gaussian Mixture Model (GMM) components are assigned to pixels and learned from the 4-channel Image frame. The energy model is defined based on the foreground and background histograms and the minimum energy represents good segmentation. The segmentation is estimated using Graph Cut, which provides tentative classification of pixels belonging to the respective classes. The above process is iterated until convergence. The formulation of above mentioned process is briefly described below. Let us consider image frame as an array $I = (I_1, ..I_n...I_N)$, which includes both R,G,B levels and depth values respectively. The segmentation is array of opacity values $\underline{\alpha} = (\alpha_1, ...\alpha_N)$ at each pixel. 0 for background and 1 for foreground. $\underline{\theta}$ is the model, which defines image foreground and background histogram distributions and it is defined as follows.

$$\underline{\theta} = h(I; \alpha), \alpha = 0, 1. \tag{6.1}$$

Given an image frame I and model $\underline{\theta}$, the segmentation task is to infer unknown opacity variables $\underline{\alpha}$. The energy function E is defined such that minimum represents good segmentation and can be formulated as follows.

$$\underline{\alpha} = \mathrm{argmin}_{\underline{\alpha}} E(\underline{\alpha}, \underline{\theta}). \tag{6.2}$$

The model parameters are represented by

$$\underline{\theta} = \left\{ \pi(\alpha, k), \mu(\alpha, k), \sum(\alpha, k), \alpha = 0, 1, k = 1...K \right\}, \tag{6.3}$$

where π is the weight, μ is the mean and $\sum$ is the covariance of 2K Gaussian components for foreground and background distributions. Smoothness factor V is defined as follows.

$$V(\alpha, I) = \gamma \sum_{m,n \in C} [\alpha_n \neq \alpha_m] \exp -\beta ||I_m - I_n||^2. \tag{6.4}$$

In [20], we used scaling function in smoothness factor thereby emphasizing the importance of depth, which is achieved using weighted L2 norm.

In fact, user initialization can be avoided by using relevant visual attention models.

6.6 Augmented Reality

Augmented Reality is widely used in many applications. For example, in [24], a list of around thirty interesting augmented reality applications has been discussed. Liu et al. [15] has used model-based video segmentation for interactive games.

6.6.1 Tourism

Depending on the tourist spot, system presents the cultural or heritage story. As stated in [27], system displays the user movement along with narration. Actually, using segmentation the virtual tourist guide can be associated along with the narration. This is one of the exciting applications adding real world experience through mobile devices.

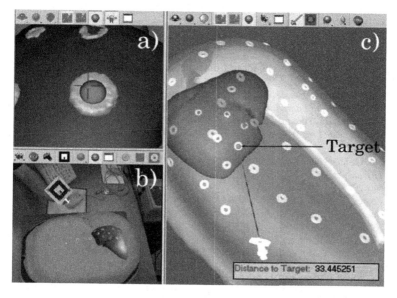

Fig. 6.10 Surgery [17]

6.6.2 *Virtual Surgery*

Augmented Reality has been used in medical fields for (a) virtual face surgery, (b) location of tumours, (c) treatment planning and (d) anatomical structure study. For example, as shown in Fig. 6.10, the technique helps surgeons with a view in transparency of their patient and by tracking surgical tools in real time [17].

6.7   Conclusions and Future Work

In this chapter, we briefly discussed segmentation applications. There is a paradigm shift from traditional segmentation to using depth, attention and prior model information in addition to color and motion-based approaches. The expected segmentation quality for a given application depends on the level of granularity and the requirement that is related to shape precision and temporal coherence of the objects. Although, there exists still significant challenge to perform robust and fully automated segmentation that fits generic tasks, a reliable solution can be achieved using suitable attention and model-based information.

Acknowledgements Our thanks to colleagues in MMV lab for their suggestions.

References

1. S. Ashar, Y. Kondratyuk, K. Elazouzi, P. Kauff and O. Schreer, Vision based skin colour segmentation of moving hands for real time applications, *Conference on Visual Media and Production*, CVMP 2004.
2. S.F. Chang, Content-based video summarization and adaptation for ubiquitous media access, *Proceedings of 12th International Conference on Image Analysis and Processing*, 2003.
3. P. L. Correia and F. Pereira, Classification of Video Segmentation Application Scenarios, *IEEE Transactions on Circuits and Systems for Video Technology*, series 5, Vol. 14, pp. 735–741, 2004.
4. C. Cui, Q. Zhang and K.N. Ngan, Multi-view Video Based Object Segmentation - A Tutorial, *ECTI Transactions on Electrical Engineering, Electronics and Communications*, Vol. 7, No. 2, August 2009, pp. 90–105.
5. K. Forbes, F. Nicolls, G. Jager and A. Voigt, Shape-from-Silhouette with Two Mirrors and an Uncalibrated Camera, *In Proceedings of the 9th European Conference on Computer Vision (ECCV)*, May 2006.
6. E.D. Gelasca, T. Ebrahimi, On Evaluating Video Object Segmentation Quality: A Perceptually Driven Objective Metric, *IEEE Journal of Selected Topics in Signal Processing*, April 2009, Vol.3 , Issue:2, pp. 319–335.
7. A. Hampapur, R. Jain and T.E. Weymouth, Production model based digital video segmentation, *Multimedia Tools and Applications*, pp. 9-46, Vol 1, 1995.
8. L. Itti and P.F. Baldi, A Principled Approach to Detecting Surprising Events in Video, *Proceedings of IEEE Conference on Computer Vision and Pattern Recognition (CVPR)*, pp. 631–637, 2005.
9. E. Izquierdo, Mohammed Ghanbari, Key components for an advanced segmentation system. *IEEE Transactions on Multimedia* 4(1), pp. 97–113, 2002.
10. S. Jabri and Z. Duric and H. Wechsler and A. Rosenfeld, Detection and Location of People in Video Images Using Adaptive Fusion of Color and Edge Information, in *International Conference on Pattern Recognition*, ICPR, pp. 627–630, 2000.
11. A. Krutz, M. Kunter, M. Drose, M. Frater, and T. Sikora, Content-Adaptive Video Coding combining Object-based coding and H.264/AVC, *EURASIP Proceedings*, 2007.
12. D. Li, H. Lu, Model based Video Segmentation, *IEEE Workshop on Signal Processing Systems*, pp. 120–129, 2000.
13. H. Li and K.N. Ngan, Automatic Video Segmentation and Tracking for Content-Based Applications, *Advances in Visual Content Analysis and Adaptation for Multimedia Communications*, 2007.
14. H. Li , K.N. Ngan, Saliency model-based face segmentation and tracking in head-and-shoulder video sequences, *Journal of Visual Communication and Image Representation*, 2008.
15. L.K. Liu, Model-based video segmentation for vision-augmented interactive games, *Proceedings of Image and Video Communications and Processing*, Vol. 3974, pp. 432–439, 2000.
16. Naeem Ramzan, Toni Zgaljic, Ebroul Izquierdo, Scalable Video Coding: Source for Future Media Internet, *Towards the Future Internet*, pp. 205–215, 2010.
17. L. Soler, S. Nicolau, J. Schmid, C. Koehl, J. Marescaux, X. Pennec and N. Ayache, Virtual Reality and Augmented Reality in Digestive Surgery, *Proceedings of the 3rd IEEE/ACM International Symposium on Mixed and Augmented Reality*, pp. 278–279, 2004.
18. J. Shen, Motion Detection in Color Image Sequence and Shadow Elimination, in *Visual Communications and Image Processing* VCIP, January 2004, pp. 731–740.
19. S.J. McKenna, S. Jabri, Z. Duric, A. Rosenfeld and H. Wechsler, Tracking Groups of People, in *Computer Vision and Image Understanding*, Vol. 80, 2000.
20. K. Vaiapury, A. Aksay, E. Izquierdo, GrabcutD: Improved Grabcut Using Depth Information, *ACM Workshop on Surreal Media and Virtual Cloning (SMVC)*, 2010.
21. K. Vaiapury, P.K. Atrey, M.S. Kankanhalli and K. Ramakrishnan, Non Identical Duplicate Video Detection using SIFT method, *Proceedings of IEE International Conference on Visual Information Engineering* VIE, 2006.

22. T. Zgaljic, N. Ramzan, M. Akram, E. Izquierdo, R. Caballero, A. Finn, H. Wang and Z. Xiong Surveillance Centric Coding, *5th International Conference on Visual Information Engineering*, VIE 2008.

23. Z. Zivkovic, M. Petkovic, R. Van Mierlo, M. van Keulen, F. van der Heijden, W. Jonker, E. Rijnierse, Two video analysis applications using foreground/background segmentation, *International Conference on Visual Information Engineering*, VIE 2003, pp. 310–313, 2003.

24. http://www.iphoneness.com/iphone-apps/best-augmented-reality-iphone-applications/ (retrieved as on 14-10-2010).

25. http://news.cnet.com/2300-1035_3-10005011.html/ (retrieved as on 14-10-2010).

26. http://news.bbc.co.uk/1/hi/sci/tech/1953770.stm/ (retrieved as on 14-10-2010).

27. http://blogs.adobe.com/jd1/archives/2005/07/augmented-touri.html/ (retrieved   as   on 14-10-2010).

28. http://jaxov.com/2009/11/track-your-parked-car-with-car-finder-iphone-app-augmented-reality/ (retrieved as on 14-10-2010).

29. http://research.microsoft.com/en-us/um/people/sbkang/3dvideodownload/ (retrieved as on 14-10-2010).

Index

K.N. Ngan and H. Li (eds.), *Video Segmentation and Its Applications*,
DOI 10.1007/978-1-4419-9482-0, © Springer Science+Business Media, LLC 2011